# Lecture Notes in Artificial Intelligence 5819

Edited by R. Goebel, J. Siekmann, and W. Wahlster

Subseries of Lecture Notes in Computer Science

T0073528

Ning Zhong   Kuncheng Li   Shengfu Lu
Lin Chen (Eds.)

# Brain Informatics

International Conference, BI 2009
Beijing, China, October 22-24, 2009
Proceedings

 Springer

Series Editors

Randy Goebel, University of Alberta, Edmonton, Canada
Jörg Siekmann, University of Saarland, Saarbrücken, Germany
Wolfgang Wahlster, DFKI and University of Saarland, Saarbrücken, Germany

Volume Editors

Ning Zhong
Department of Life Science and Informatics, Maebashi Institute of Technology, Japan
E-mail: zhong@maebashi-it.ac.jp

Kuncheng Li
Radiology, Xuanwu Hospital of Capital Medical University, Beijing, China
E-mail: likuncheng1955@yahoo.com.cn

Shengfu Lu
The International WIC Institute, Beijing University of Technology, China
E-mail: lusf@bjut.edu.cn

Lin Chen
Institute of Biophysics, Chinese Academy of Sciences, Beijing, China
E-mail: cl@cogsci.ibp.ac.cn

Library of Congress Control Number: 2009936006

CR Subject Classification (1998): I.2, H.5.1-3, I.2.8, K.3, J.3, I.4-5, J.1

LNCS Sublibrary: SL 7 – Artificial Intelligence

| | |
|---|---|
| ISSN | 0302-9743 |
| ISBN-10 | 3-642-04953-2 Springer Berlin Heidelberg New York |
| ISBN-13 | 978-3-642-04953-8 Springer Berlin Heidelberg New York |

springer.com

© Springer-Verlag Berlin Heidelberg 2009
Printed in Germany

Typesetting: Camera-ready by author, data conversion by Scientific Publishing Services, Chennai, India
Printed on acid-free paper     SPIN: 12772292     06/3180     5 4 3 2 1 0

# Preface

This volume contains the papers selected for presentation at The 2009 International Conference on Brain Informatics (BI 2009) held at Beijing University of Technology, China, on October 22–24, 2009. It was organized by the Web Intelligence Consortium (WIC) and IEEE Computational Intelligence Society Task Force on Brain Informatics (IEEE TF-BI). The conference was held jointly with The 2009 International Conference on Active Media Technology (AMT 2009).

Brain informatics (BI) has emerged as an interdisciplinary research field that focuses on studying the mechanisms underlying the human information processing system (HIPS). It investigates the essential functions of the brain, ranging from perception to thinking, and encompassing such areas as multi-perception, attention, memory, language, computation, heuristic search, reasoning, planning, decision-making, problem-solving, learning, discovery, and creativity. The goal of BI is to develop and demonstrate a systematic approach to achieving an integrated understanding of both macroscopic and microscopic level working principles of the brain, by means of experimental, computational, and cognitive neuroscience studies, as well as utilizing advanced Web Intelligence (WI) centric information technologies. BI represents a potentially revolutionary shift in the way that research is undertaken. It attempts to capture new forms of collaborative and interdisciplinary work. Following this vision, new kinds of BI methods and global research communities will emerge, through infrastructure on the wisdom Web and knowledge grids that enables high speed and distributed, large-scale analysis and computations, and radically new ways of sharing data/knowledge.

BI 2009 was the first conference specifically dedicated to interdisciplinary research in brain informatics. It provided an international forum to bring together researchers and practitioners from diverse fields, such as computer science, information technology, artificial intelligence, Web intelligence, cognitive science, neuroscience, medical science, life science, economics, data mining, data science and knowledge science, intelligent agent technology, human computer interaction, complex systems, and systems science, to present the state-of-the-art in the development of brain informatics, to explore the main research problems in BI that lie in the interplay between the studies of the human brain and the research of informatics. On the one hand, one models and characterizes the functions of the human brain based on the notions of information processing systems. WI centric information technologies are applied to support brain science studies. For instance, the wisdom Web, knowledge grids, and cloud computing enable high-speed, large-scale analysis, simulation, and computation as well as new ways of sharing research data and scientific discoveries. On the other hand, informatics-enabled brain studies, e.g., based on fMRI, EEG, and MEG, significantly broaden the spectrum of theories and models of brain sciences and offer new insights into the development of human-level intelligence towards brain-inspired wisdom Web computing.

We wish to express our gratitude to all members of the Conference Committee for their instrumental and unfailing support. BI 2009 had a very exciting program with a number of features, ranging from keynote talks, special sessions, technical sessions, posters, workshops, and social programs. All of this work would not have been possible without the generous dedication of the Program Committee members and the external reviewers in reviewing the papers submitted to BI 2009, of our keynote speakers, John Anderson of Carnegie Mellon University, Jeffrey M. Bradshaw of Florida Institute for Human and Machine Cognition, Frank van Harmelen of Vrije Universiteit Amsterdam, Lynne Reder of Carnegie Mellon University, Zhongzhi Shi of the Chinese Academy of Sciences, and Zhi-Hua Zhou of Nanjing University, and invited speakers in the special session on Information Processing Meets Brain Sciences, Zhaoping Li of University College London, Setsuo Ohsuga of the University of Tokyo, Bin Hu of Birmingham City University and Lanzhou Unviersity, Tianzi Jiang of the Chinese Academy of Sciences, Yulin Qin of Beijing University of Technology and Carnegie Mellon University, in preparing and presenting their very stimulating talks. We thank them for their strong support.

BI 2009 could not have taken place without the great team effort of the Local Organizing Committee and the support of the International WIC Institute, Beijing University of Technology. Our special thanks go to Boyuan Fan, Ze Zhang, Zhenyang Lu, Pu Wang, and Jianwu Yang for their enormous efforts in planning and arranging the logistics of the conference from registration/payment handling, venue preparation, accommodation booking, to banquet/social program organization. We would like to thank Shuai Huang, Jiajin Huang, Jian Yang, and Juzhen Dong of the conference support team at the International WIC Institute (WICI), the Knowledge Information Systems Laboratory, Maebashi Institute of Technology, and Web Intelligence Laboratory, Inc. for their dedication and hard work. We are very grateful to the BI 2009 corporate sponsors: Beijing University of Technology (BJUT), Beijing Municipal Lab of Brain Informatics, Chinese Society of Radiology, National Natural Science Foundation of China (NSFC), State Administration of Foreign Experts Affairs, Shanghai Psytech Electronic Technology Co. Ltd, Shenzhen Hanix United, Inc. (Beijing Branch), Beijing JinShangQi Net System Integration Co. Ltd, and Springer Lecture Notes in Computer Science (LNCS/LNAI) for their generous support. Last but not the least, we thank Alfred Hofmann of Springer for his help in coordinating the publication of this special volume in an emerging and interdisciplinary research field.

August 2009                                                        Ning Zhong
                                                                   Kuncheng Li
                                                                   Shengfu Lu
                                                                   Lin Chen

# Conference Organization

## Conference General Chairs

| | |
|---|---|
| Lin Chen | Chinese Academy of Sciences, China |
| Setsuo Ohsuga | University of Tokyo, Japan |

## Program Chairs

| | |
|---|---|
| Ning Zhong | International WIC Institute, Beijing University of Technology, China Maebashi Institute of Technology, Japan |
| Kuncheng Li | Xuanwu Hospital, Capital Medical University, China |

## Workshop Chairs

| | |
|---|---|
| Yiyu Yao | International WIC Institute, Beijing University of Technology, China University of Regina, Canada |
| Runhe Huang | Hosei University, Japan |

## Organizing Chairs

| | |
|---|---|
| Shengfu Lu | International WIC Institute, Beijing University of Technology, China |
| Xunming Ji | Xuanwu Hospital, Capital Medical University, China |

## Publicity Chairs

| | |
|---|---|
| Jian Yang | International WIC Institute, Beijing University of Technology, China |
| Jiajin Huang | International WIC Institute, Beijing University of Technology, China |

## WIC Co-chairs/Directors

| | |
|---|---|
| Ning Zhong | Maebashi Institute of Technology, Japan |
| Jiming Liu | Hong Kong Baptist University, HK |

## IEEE TF-BI Chair

Ning Zhong               Maebashi Institute of Technology, Japan

## WIC Advisory Board

Edward A. Feigenbaum    Stanford University, USA
Setsuo Ohsuga          University of Tokyo, Japan
Benjamin Wah           University of Illinois, Urbana-Champaign,
                           USA
Philip Yu              University of Illinois, Chicago, USA
L.A. Zadeh             University of California, Berkeley, USA

## WIC Technical Committee

Jeffrey Bradshaw       UWF/Institute for Human and Machine
                           Cognition, USA
Nick Cercone           York University, Canada
Dieter Fensel          University of Innsbruck, Austria
Georg Gottlob          Oxford University, UK
Lakhmi Jain            University of South Australia, Australia
Jianchang Mao          Yahoo! Inc., USA
Pierre
   Morizet-Mahoudeaux  Compiègne University of Technology, France
Hiroshi Motoda         Osaka University, Japan
Toyoaki Nishida        Kyoto University, Japan
Andrzej Skowron        Warsaw University, Poland
Jinglong Wu            Okayama University, Japan
Xindong Wu             University of Vermont, USA
Yiyu Yao               University of Regina, Canada

## Program Committee

Yiuming Cheung         Hong Kong Baptist University, HK
Xiaocong Fan           The Pennsylvania State University, USA
Mohand-Said Hacid      Université Claude Bernard Lyon 1, France
Kazuyuki Imamura       Maebashi Institute of Technology, Japan
Akrivi Katifori        University of Athens, Greece
Peipeng Liang          Beijing University of Technology, China
Pawan Lingras          Saint Mary's University, Canada
Duoqian Miao           Tongji University, China
Mariofanna Milanova    University of Arkansas at Little Rock, USA
Sankar Kumar Pal       Indian Statistical Institute, India
Hideyuki Sawada        Kagawa University, Japan

# Table of Contents

## Information Technologies for the Management and Use of Brain Data

## Cognition-inspired Applications

# Using Neural Imaging to Inform the Instruction of Mathematics

John Anderson

Department of Psychology
Carnegie Mellon University, USA
ja+@cmu.edu

I will describe research using fMRI to track the learning of mathematics with a computer-based algebra tutor. I will describe the methodological challenges in studying such a complex task and how we use cognitive models in the ACT-R architecture to interpret imaging data. I wll also describe how we can use the imaging data to identify mental states as the student is engaged in algebraic problems solving.

N. Zhong et al. (Eds.): BI 2009, LNAI 5819, p. 1, 2009.

# Distributed Human-Machine Systems: Progress and Prospects

Jeffrey M. Bradshaw

Florida Institute for Human and Machine Cognition, USA
jbradshaw@ihmc.us

Advances in neurophysiological and cognitive science research have fueled a surge of research aimed at more effectively combining human and machine capabilities. In this talk we will give and overview of progress and prospects for four current thrusts of technology development resulting from this research: brain-machine interfaces, robotic prostheses and orthotics, cognitive and sensory prostheses, and software and robotic assistants. Following the overview, we will highlight the unprecedented social ethics issues that arise in the design and deployment of such technologies, and how they might be responsibly considered and addressed.

N. Zhong et al. (Eds.): BI 2009, LNAI 5819, p. 2, 2009.
© Springer-Verlag Berlin Heidelberg 2009

# Large Scale Reasoning on the Semantic Web: What to Do When Success Is Becoming a Problem

Frank van Harmelen

AI Department
Vrije Universiteit Amsterdam, The Netherland
Frank.van.Harmelen@cs.vu.nl

In recent years, the Semantic Web has seen rapid growth in size (many billions of facts and rules are now available) and increasing adoption in many sectors (government, publishing industry, media). This success has brought with it a whole new set of problems: storage, querying and reasoning with billions of facts and rules that are distributed across different locations. The Large Knowledge Collider (LarKC) is providing an infrastructure to solve such problems. LarKC exploits parallelisation, distribution and approximation to enable Semantic Web reasoning at arbitrary scale. In this presentation we will describe the architecture and implementation of the Large Knowledge Collider, we will give data on its current performance, and we will describe a number of use-cases that are deploying LarKC.

N. Zhong et al. (Eds.): BI 2009, LNAI 5819, p. 3, 2009.
© Springer-Verlag Berlin Heidelberg 2009

# How Midazolam Can Help Us Understand Human Memory: 3 Illustrations and a Proposal for a New Methodology

Lynne Reder

Department of Psychology
Carnegie Mellon University, USA
reder@cmu.edu

Midazolam is a benzodiazepine commonly used as an anxiolytic in surgery. A useful attribute of this drug is that it creates temporary, reversible, anterograde amnesia. Studies involving healthy subjects given midazolam in one session and saline in another, in a double-blind, cross-over design, provide insights into memory function. Several experiments will be described to illustrate the potential of studying subjects with transient anterograde amnesia. This talk will also outline how this drug can be used in combination with fMRI to provide more insights about brain functioning than either method in isolation.

N. Zhong et al. (Eds.): BI 2009, LNAI 5819, p. 4, 2009.
© Springer-Verlag Berlin Heidelberg 2009

# Research on Brain-Like Computer

Zhongzhi Shi

Key Laboratory of Intelligent Information Processing
Institute of Computing Technology, Chinese Academy of Sciences
Beijing 100190, China
shizz@ics.ict.ac.cn

After more than 60 years of development, the operation speed of computer is up to several hundred thousand billion ($10^{14}$ ) times, but its intelligence level is extremely low. Studying machine which combines high performance and the people's high intelligence together becomes the effective way with high capacity and efficiency of exploring information processing. It will bring the important impetus to economic and social sustainable development, promotion of the information industry and so on to make breakthrough in the research of brain-like computer.

Mind is all mankind's spiritual activities, including emotion, will, perception, consciousness, representation, learning, memory, thinking, intuition, etc. Mind model is for explaining what individuals operate in the cognitive process for some thing in the real world. It is the internal sign or representation for external realistic world. If the neural network is a hardware of the brain system, then the mind model is the software of the brain system. The key idea in cognitive computing is to set up the mind model of the brain system, and then building brain-like computer in engineering through structure, dynamics, function and behavioral reverse engineering of the brain. This talk will introduce the research progress of brain-like computer, mainly containing intelligence science, mind modesl, neural columns, architecture of brain-like computers.

Intelligence Science is an interdisciplinary subject which dedicates to joint research on basic theory and technology of intelligence by brain science, cognitive science, artificial intelligence and others. Brain science explores the essence of brain, research on the principle and model of natural intelligence in molecular, cell and behavior level. Cognitive science studies human mental activity, such as perception, learning, memory, thinking, consciousness etc. In order to implement machine intelligence, artificial intelligence attempts simulation, extension and expansion of human intelligence using artificial methodology and technology. Research scientists coming from above three disciplines work together to explore new concept, new theory, new methodology. Intelligence science is a essential way to reach the human-level intelligence and point out the basic priciples for brain-like computer.

N. Zhong et al. (Eds.): BI 2009, LNAI 5819, p. 5, 2009.

# A Framework for Machine Learning with Ambiguous Objects

Zhi-Hua Zhou

National Key Laboratory for Novel Software Technology
Nanjing University, Nanjing 210093, China
zhouzh@nju.edu.cn

Machine learning tries to improve the performance of the system automatically by learning from experiences, e.g., objects or events given to the system as training samples. Generally, each object is represented by an instance (or feature vector) and is associated with a class label indicating the semantic meaning of that object. For ambiguous objects which have multiple semantic meanings, traditional machine learning frameworks may be less powerful. This talk will introduce a new framework for machine learning with ambiguous objects.

N. Zhong et al. (Eds.): BI 2009, LNAI 5819, p. 6, 2009.

# Data Compression and Data Selection in Human Vision

Zhaoping Li

University College London, UK
Z.Li@cs.ucl.ac.uk

The raw visual inputs give many megabytes of data per second to human eyes, but there are neural and cognitive bottle necks in human vision such that only a few dozens of bits of data per second are eventually perceived. I will talk about two processes in the early stages of visual pathway to reduce data rate. One is to compress data such that as much information as possible is transmitted to the brain given a limited information channel capacity in the visual pathway. The other is to select, through visual attention, a fraction of the transmitted information for further detailed processing, and the data not selected are ignored. See http://www.cs.ucl.ac.uk/staff/Zhaoping.Li/prints/ZhaopingNReview2006.pdf for details.

N. Zhong et al. (Eds.): BI 2009, LNAI 5819, p. 7, 2009.
© Springer-Verlag Berlin Heidelberg 2009

# Do Brain Networks Correlate with Intelligence?

Tianzi Jiang

LIAMA Center for Computational Medicine
National Laboratory of Pattern Recognition
Institute of Automation, Chinese Academy of Sciences
Beijing 100080, P.R. China
jiangtz@nlpr.ia.ac.cn

Intuitively, higher intelligence might be assumed to correspond to more efficient information transfer in the brain, but no direct evidence has been reported from the perspective of brain networks. In this lecture, we first give a brief introduction about the basic concepts of brain networks from different scales and classified ways. In the second part, we present the advance on how functional brain networks correlate with intelligence. We focus on the evidence obtained with functional magnetic resonance imaging (fMRI) in the rest state. In the third part of this lecture, we will discuss how individual differences in intelligence are associated with brain structural organization, and in particular that higher scores on intelligence tests are related to greater global efficiency of the brain anatomical network. We focus on the evidence obtained with diffusion tensor imaging (DTI), a type of magnetic resonance imaging. In the fourth part, we discuss the genetic basis of intelligence-related brain networks. We try to address the issue on how intelligence-related genes influence intelligence-related neuronal systems. The evidence based on fMRI and DTI are presented. Finally, the future directions in this field will be presented.

N. Zhong et al. (Eds.): BI 2009, LNAI 5819, p. 8, 2009.
© Springer-Verlag Berlin Heidelberg 2009

# How Were Intelligence and Language Created in Human Brain

Setsuo Ohsuga

University of Tokyo, Japan
ohsuga@s01.itscom.net

The final objective of this talk is to make clear the origin of language and intelligence. Since intelligence depends on language, origin of language rather than intelligence is mainly discussed.

First of all a problem immanent in the current understanding of language is discussed. The structure and the meaning of languages were analyzed by many philosophers such as F. Saussure, F.L.G. Frege, B.A.W. Russell, L.J.J. Wittgenstein, N. Chomsky and the others, and the characteristics of language were made almost clear by them. What is to be noted in their discussions is that language is thought originated from a latent potentiality of human being. F. Saussure named it language in contrast to real languages, named langue by him, that are generated from language. Chomsky asserts Universal Grammar as the origin of language and that it is biologically inherent to human being. But they do not explain anything about what is this latent potentiality or how language was created. If human being came out of an evolutional process of living things, then language had to be created at some point in this process. There was nothing before that. How primitive human being could create such complex existence as language which many philosophers with highest intelligence such as listed above had interested in. This is a big mystery.

In order to have an insight to this problem it is necessary to study the language along the time axis, from very ancient time when primitive human being have no language to today, and investigate the possibility of language having been created. In this talk, the process of language development is classified into the following four stages,

1. Before language - forming primitive concept
2. Primitive language - from holistic language to compositional language
3. Social language and symbolic language
4. Visual language - invention of letters and high order expressions

Among these stages, this article discuses especially the first two stages. The major interest is to find the way a basic structure of language that corresponds to Universal Grammar of Chomsky could be produced from the biological functions of human brains.

The first point to be discussed is the way primitive human being created concepts. The concepts had to be made before language because language is a way to send existing concepts to the others.

In this talk it is assumed that predicate logic can be the Universal Grammar because it can be a framework of every language. Then, it is shown that the production rules of well formed formula (wff) of predicate logic can be realized by

N. Zhong et al. (Eds.): BI 2009, LNAI 5819, pp. 9–10, 2009.

neural network with such characteristics as, new neuron is born, neuron grows by extending its dendrite, a neurons makes links to the others by touching its dendrite to the axon of the others resulting in a neural-network, every neuron learns from stimulus from external world and the success or the failure of its response. If further more it is shown that a neuron could behave like a logical inference by learning, then these biological characteristics can be a latent Universal Grammar and the created neural networks represent a language expression. The difference between ordinary operation of neural network like recognition and this logical operation is made clear.

By showing that these concepts can be connected to the vocal organs, and also to sensor organs by means also of purely biological ways, a possibility of originating language could be explained.

This was the simplest holistic language. It was shown that it could develop to social language and to compositional language in the same framework. After some time this language was symbolized. Symbolization was a next big evolutional stage of language. It depends deeply on biological capability of memorizing. But this function is not made clear yet. In order to analyze the process of symbolization we must wait the researches from neuroscience and brain science on this matter.

# Affective Learning with an EEG Approach

Bin Hu

School of Information Science and Engineering
Lanzhou University, China, and
Department of Computing, Birmingham City University, UK
binhu@lzu.edu.cn

People's moods heavily influence their way of communicating and their acting and productivity, which also plays a crucial role in learning process. Affective learning is an important aspect of education. Emotions of learners need to be recognized and interpreted so as to motivate learners and deepen their learning, and this is a prerequisite in e-learning. Normally, affective learning has been investigating some technologies to understand learners' emotions through detecting their face, voice and eyes motion, etc.

Our research focuses on how to enhance interactive learning between learners and tutors, and understand learners' emotions through an EEG approach. We have developed an e-learning environment and recorded EEG signals of learners while they are surfing in the website. We presented an ontology based model for analyzing learners' alpha wave (a component of EEG signals) to infer the meaning of its representation, then to understand learners' emotions in learning process. The outcomes of the research can contribute to evaluation of e-learning systems and deepen understanding of learners' emotions in learning process.

N. Zhong et al. (Eds.): BI 2009, LNAI 5819, p. 11, 2009.

# Some Web Intelligence Oriented Brain Informatics Studies

Yulin Qin

The International WIC Institute
Beijing University of Technology, China, and
Department of Psychology, Carnegie Mellon University, USA
yulinqin@gmail.com

With the advancement both in the Web (e.g., semantic Web and human-level wisdom-Web computing) and in Brain Informatics (BI) (e.g., advanced information technologies for brain science and non-invasive neuroimaging technologies, such as functional magnetic resonance imaging (fMRI)), several lines of BI research have been developed directly or indirectly related to Web Intelligence (WI). Some of them can be treated as the extension to the Web research of the tradition BI research, such as computational cognitive modeling like ACT-R. ACT-R is a theory and model of computational cognitive architecture which consists of functional modules, such as declarative knowledge module, procedural knowledge module, goal module and input (visual, aural), output (motor, verbal) modules. Information can be proposed parallel inside and among the modules, but has to be sequentially if it needs procedural module to coordinate the behavior across modules. At the International WIC Institute (WICI), we are trying to introduce this kind of architecture and the mechanism of activation of the units in declarative knowledge module into our Web information system. Based on or related to ACT-R, theories and models that are with very close relation to WI have also been developed, such as threaded cognition for concurrent multitasking, cognitive agents, human-Web interaction (e.g., SNIT-ACT (Scent-based navigation and information foraging in the ACT cognitive architecture). At the WICI, we are also working on the user behavior and reasoning on the Web by eye-tracker and fMRI. Some of other BI studies, however, have been developed directly by the requirement of WI research. For example, to meet the requirement of the development of Granular Reasoning (GrR) technologies in WI research, people at the WICI have been checking how human can perceive the real world under many levels of granularity (i.e., abstraction) and can also easily switch among granularities. By focusing on different levels of granularity, one can obtain different levels of knowledge, as well as in-depth understanding of the inherent knowledge structure. The interaction between human intelligence inspired WI methodology research and WI stimulated BI principle research will benefit both BI and WI researches greatly.

N. Zhong et al. (Eds.): BI 2009, LNAI 5819, p. 12, 2009.

# Modelling the Reciprocal Interaction between Believing and Feeling from a Neurological Perspective

Zulfiqar A. Memon[1,2] and Jan Treur[1]

[1] VU University Amsterdam, Department of Artificial Intelligence
De Boelelaan 1081, 1081 HV Amsterdam
[2] Sukkur Institute of Business Administration (Sukkur IBA),
Air Port Road Sukkur, Sindh, Pakistan
{zamemon,treur}@few.vu.nl
http://www.few.vu.nl/~{zamemon,treur}

**Abstract.** By adopting neurological theories on the role of emotions and feelings, an agent model is introduced incorporating the reciprocal interaction between believing and feeling. The model describes how the strength of a belief may not only depend on information obtained, but also on the emotional responses on the belief. For feeling emotions a recursive body loop is assumed. The model introduces a second feedback loop for the interaction between feeling and belief. The strength of a belief and of the feeling both result from the converging dynamic pattern modelled by the combination of the two loops. For some specific cases it is described, for example, how for certain personal characteristics an optimistic world view emerges, or, for other characteristics, a pessimistic world view.

## 1 Introduction

Already during the process that they are generated beliefs trigger emotional responses that result in certain feelings. However, the process of generation of a belief is not fully independent of such associated feelings. In a reciprocal manner, the generated feelings may also have a strengthening or weakening effect on the belief during this process. Empirical work such as described in, for example, (Eich, Kihlstrom, Bower, Forgas, and Niedenthal, 2000; Forgas, Laham, and Vargas, 2005; Forgas, Goldenberg, and Unkelbach, 2009; Niedenthal, 2007; Schooler and Eich, 2000; Winkielman, Niedenthal, and Oberman, 2009), reports such types of effects of emotions on beliefs, but does not relate them to neurological findings or theories. In this paper, adopting neurological theories on emotion and feeling, a computational dynamic agent model is introduced that models this reciprocal interaction between feeling and believing. The computational model, which is based on neurological theories on the embodiement of emotions as described, for example, in (Damasio, 1994, 1996, 1999, 2004; Winkielman, Niedenthal, and Oberman, 2009)'s, describes how the generation of a belief may not only depend on an (external) informational source, but also takes into account how the belief triggers an emotional response that leads to a certain feeling. More specifically, in accordance with, for example (Damasio, 1999, 2004), for feeling the emotion associated to a belief a converging recursive body loop is assumed. A

N. Zhong et al. (Eds.): BI 2009, LNAI 5819, pp. 13–24, 2009.

second converging feedback loop introduced in the model, inspired the Somatic Marker Hypothesis (Damasio, 1994, 1996), involves the interaction back from the feeling to the belief. Thus a combination of two loops is obtained, where connection strengths within these loops in principle are person-specific. Depending on these personal characteristics, from a dynamic interaction within and between the two loops, an equilibrium is reached for both the strength of the belief and of the feeling.

To illustrate the model, the following example scenario is used. A person is parking his car for a short time at a place where this is not allowed. When he comes back, from some distance he observes that a small paper is attached at the front window of the car. He starts to generate the belief that the paper represents a charge to be paid. This belief generates a negative feeling, which has an impact on the belief by strengthening it. Coming closer, some contours of the type of paper that is attached become visible. As these are not clearly recognized as often occurring for a charge, the person starts to generate a second belief, namely that it concerns an advertising of a special offer. This belief generates a positive feeling which has an impact on the latter belief by strengthening it.

In this paper, first in Section 2 Damasio's theory on the generation of feelings based on a body loop is briefly introduced. Moreover, the second loop is introduced, the one between feeling and belief. In Section 3 the model is described in detail. Section 4 presents some simulation results. In Section 5 a mathematical analysis of the equilibria of the model is presented. Finally, Section 6 is a discussion.

## 2   From Believing to Feeling and Vice Versa

In this section the interaction between believing and feeling is discussed in some more detail from a neurological perspective, in both directions: from believing to feeling, and from feeling to believing.

### 2.1   From Believing to Feeling

As any mental state in a person, a belief state induces emotions felt within this person, as described by Damasio (1999, 2004); for example:

> 'Even when we somewhat misuse the notion of feeling – as in "I feel I am right about this" or "I feel I cannot agree with you" – we are referring, at least vaguely, to the feeling that accompanies the idea of believing a certain fact or endorsing a certain view. This is because believing and endorsing *cause* a certain emotion to happen. As far as I can fathom, few if any exceptions of any object or event, actually present or recalled from memory, are ever neutral in emotional terms. Through either innate design or by learning, we react to most, perhaps all, objects with emotions, however weak, and subsequent feelings, however feeble.' (Damasio, 2004, p. 93)

In some more detail, emotion generation via a body loop roughly proceeds according to the following causal chain; see Damasio (1999, 2004):

belief  → preparation for the induced bodily response      →
induced bodily response        → sensing the induced bodily response →
sensory representation of the induced bodily response → induced feeling

As a variation, an 'as if body loop' uses a direct causal relation

preparation for the induced bodily response $\rightarrow$
sensory representation of the induced bodily response

as a shortcut in the causal chain. The body loop (or as if body loop) is extended to a recursive body loop (or recursive as if body loop) by assuming that the preparation of the bodily response is also affected by the state of feeling the emotion:

feeling $\rightarrow$ preparation for the bodily response

as an additional causal relation. Such recursiveness is also assumed by Damasio (2004), as he notices that what is felt by sensing is actually a body state which is an internal object, under control of the person:

> 'The brain has a direct means to respond to the object as feelings unfold because the object at the origin is inside the body, rather than external to it. The brain can act directly on the very object it is perceiving. It can do so by modifying the state of the object, or by altering the transmission of signals from it. The object at the origin on the one hand, and the brain map of that object on the other, can influence each other in a sort of reverberative process that is not to be found, for example, in the perception of an external object.' (…)
> 'In other words, feelings are not a passive perception or a flash in time, especially not in the case of feelings of joy and sorrow. For a while after an occasion of such feelings begins – for seconds or for minutes – there is a dynamic engagement of the body, almost certainly in a repeated fashion, and a subsequent dynamic variation of the perception. We perceive a series of transitions. We sense an interplay, a give and take.' (Damasio, 2004, pp. 91-92)

Thus the obtained model is based on reciprocal causation relations between emotion felt and body states, as roughly shown in Figure 1.

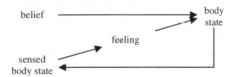

**Fig. 1.** Body loop induced by a belief

Within the model presented in this paper both the bodily response and the feeling are assigned a level or gradation, expressed by a number, which is assumed dynamic; for example, the strength of a smile and the extent of happiness. The causal cycle is modelled as a positive feedback loop, triggered by a mental state and converging to a certain level of feeling and body state. Here in each round of the cycle the next body state has a level that is affected by both the mental state and the level of the feeling state, and the next level of the feeling is based on the level of the body state.

## 2.2   From Feeling to Believing

In an idealised rational agent the generation of beliefs might only depend on informational sources and be fully independent from non-informational aspects such as emotions. However, in real life persons may, for example, have a more optimistic or pessimistic character and affect their beliefs in the sense that an optimist person strengthens beliefs that have a positive feeling associated and a pessimistic person strengthens beliefs with a negative associated feeling. Thus the strengths of beliefs may depend on non-informational aspects of mental processes and related personal characteristics. To model this for the case of feelings a causal relation

feeling → belief

can be added. This introduces a second recursive loop, as shown in Figure 2.

**Fig. 2.** The two recursive loops related to a belief

From a neurological perspective the existence of a connection from feeling to belief may be considered plausible, as neurons involved in the belief and in the associated feeling will often be activated simultaneously. Therefore such a connection from feeling to belief may be developed based on a general Hebbian learning mechanism (Hebb, 1949; Bi and Poo, 2001) that strengthens connections between neurons that are activated simultaneously, similar to what has been proposed for the emergence of mirror neurons; e.g., (Keysers and Perrett, 2004; Keysers and Gazzola, 2009).

Another type of support for a connection from feeling to belief can be found in Damasio's Somatic Marker Hypothesis; cf. (Damasio, 1994, 1996; Bechara and Damasio, 2004; Damasio, 2004). This is a theory on decision making which provides a central role to emotions felt. Each decision option induces (via an emotional response) a feeling which is used to mark the option. For example, when a negative somatic marker is linked to a particular option, it provides a negative feeling for that option. Similarly, a positive somatic marker provides a positive feeling for that option. Damasio describes the use of somatic markers in the following way:

> 'the somatic marker (..) forces attention on the negative outcome to which a given action may lead, and functions as an automated alarm signal which says: Beware of danger ahead if you choose the option which leads to this outcome. The signal may lead you to reject, *immediately*, the negative course of action and thus make you choose among other alternatives. (…) When a positive somatic marker is juxtaposed instead, it becomes a beacon of incentive. (…) on occasion somatic markers may operate covertly (without coming to consciousness) and may utilize an 'as-if-loop'.' (Damasio, 1994, p. 173-174)

Usually the Somatic Marker Hypothesis is applied to provide endorsements or valuations for options for a person's actions. However, it may be considered plausible that such a mechanism is applicable to valuations of internal states such as beliefs as well.

## 3  The Detailed Agent Model for Believing and Feeling

Informally described theories in scientific disciplines, for example, in biological or neurological contexts, often are formulated in terms of causal relationships or in terms of dynamical systems. To adequately formalise such a theory the hybrid dynamic modelling language LEADSTO has been developed that subsumes qualitative and quantitative causal relationships, and dynamical systems; cf. (Bosse, Jonker, Meij and Treur, 2007). This language has been proven successful in a number of contexts, varying from biochemical processes that make up the dynamics of cell behaviour (cf. Jonker, Snoep, Treur, Westerhoff, Wijngaards, 2008) to neurological and cognitive processes (e.g., Bosse, Jonker, Los, Torre, and Treur, 2007; Bosse, Jonker, and Treur, 2007, 2008). Within LEADSTO the temporal relation a → b denotes that when a state property a occurs, then after a certain time delay (which for each relation instance can be specified as any positive real number), state property b will occur. In LEADSTO both logical and numerical calculations can be specified in an integrated manner, and a dedicated software environment is available to support specification and simulation.

An overview of the agent model for believing and feeling is depicted in Figure 3. This picture also shows representations from the detailed specifications explained below. However, note that the precise numerical relations between the indicated variables V shown are not expressed in this picture, but in the detailed specifications of properties below, which are labeled by LP1 to LP9 as also shown in the picture.

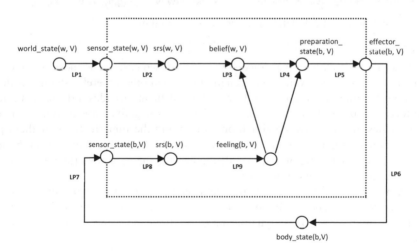

**Fig. 3.** Overview of the agent model

The detailed specification (both informally and formally) of the agent model is presented below. Here capitals are used for (assumed universally quantified) variables. First the part is presented that describes the basic mechanisms to generate a belief state and the associated feeling. The first dynamic property addresses how properties of the world state can be sensed.

**LP1  Sensing a world state**
If      world state property W occurs of strength $V$
then    a sensor state for W of strength $V$ will occur.
  world_state(W, V) $\rightarrow$ sensor_state(W, V)

For the example scenario this dynamic property is used by the agent to observe both the paper attached looking like a charge and the paper type looking like an offer; to this end the variable W is instantiated by charge and offer. From the sensor states, sensory representations are generated according to the dynamic property LP2. Note that also here for the example the variable W is instantiated as indicated.

**LP2  Generating a sensory representation for a sensed world state**
If      a sensor state for world state property W with level $V$ occurs,
then    a sensory representation for W with level $V$ will occur.
  sensor_state(W, V) $\rightarrow$ srs(W, V)

Next the dynamic property for the process for belief generation is described, where both the sensory representation and the feeling play their role. This specifies part of the loop between belief and feeling. The resulting level for the belief is calculated based on a function $g(\beta, V_1, V_2)$ of the original levels.

**LP3  Generating a belief state for a feeling and a sensory representation**
If      a sensory representation for w with level $V_1$ occurs,
  and   the associated feeling of b with level $V_2$ occurs
  and   the belief for w has level $V_3$
  and   $\beta_1$ is the person's orientation for believing
  and   $\gamma_1$ is the person's flexibility for beliefs
then    a belief for w with level $V_3 + \gamma_1 (g(\beta_1, V_1, V_2)-V_3) \Delta t$ will occur.
  srs(w, V$_1$) & feeling(b, V$_2$) & belief(w, V$_3$) $\rightarrow$ belief(w, V$_3$ + $\gamma_1$ (g($\beta_1$, V$_1$,V$_2$) - V$_3$) $\Delta$t)

For the function $g(\beta, V_1, V_2)$ the following has been taken:

$$g(\beta, V_1, V_2) = \beta(1-(1-V_1)(1-V_2)) + (1-\beta)V_1 V_2$$

Note that this formula describes a weighted sum of two cases. The most positive case considers the two source values as strengthening each other, thereby staying under $1$: combining the imperfection rates $1-V_1$ and $1-V_2$ of them provides a decreased rate of imperfection expressed by: $1-(1-V_1)(1-V_2)$. The most negative case considers the two source values in a negative combination: combining the imperfections of them provides an increased imperfection. This is expressed by $V_1 V_2$. The factor $\beta$ can be used to model the characteristic of a person that expresses the person's orientation (from $0$ as most negative to $1$ as most positive).

Dynamic property LP4 describes the emotional response to a belief in the form of the preparation for a specific bodily reaction. This specifies part of the loop between feeling and body state. This dynamic property uses the same combination model based on $g(\beta, V_1, V_2)$ as above.

**LP4  From belief and feeling to preparation of a body state**
If      belief w with level $V_1$ occurs
  and   feeling the associated body state b has level $V_2$
  and   the preparation state for b has level $V_3$
  and   $\beta_2$ is the person's orientation for emotional response
  and   $\gamma_2$ is the person's flexibility for bodily responses
then    preparation state for body state b will occur with level $V_3 + \gamma_2 (g(\beta_2, V_1, V_2)-V_3) \Delta t$.
  belief(w, V$_1$) & feeling(b, V$_2$) & preparation_state(b, V$_3$)
    $\rightarrow$ preparation_state(b, V$_3$+$\gamma_2$ (g($\beta_2$, V$_1$, V$_2$)-V$_3$) $\Delta$t)

Dynamic properties LP5 to LP9 describe the body loop.

**LP5  From preparation to effector state for body modification**
If       preparation state for body state B occurs with level $V$,
then    the effector state for body state B with level $V$ will occur.
  preparation_state(B, V)  →  effector_state(B, V)

**LP6  From effector state to modified body state**
If       the effector state for body state B with level $V$ occurs,
then    the body state B with level $V$ will occur.
  effector_state(B, V)  →  body_state(B, V)

**LP7  Sensing a body state**
If       body state B with level $V$ occurs,
then    this body state B with level $V$ will be sensed.
  body_state(B, V)  →  sensor_state(B, V)

**LP8  Generating a sensory representation of a body state**
If       body state B with level $V$ is sensed,
then    a sensory representation for body state B with level $V$ will occur.
  sensor_state(B, V)  →  srs(B, V)

**LP9  From sensory representation of body state to feeling**
If       a sensory representation for body state B with level $V$ occurs,
then    B is felt with level $V$.
  srs(B, V)  →  feeling(B, V)

Alternatively, dynamic properties LP5 to LP8 can also be replaced by one dynamic property LP10 describing an as if body loop as follows.

**LP10  From preparation to sensed body state**
If       preparation state for body state B occurs with level $V$,
then    the effector state for body state B with level $V$ will occur.
  preparation_state(B, V)  →  srs(B, V)

# 4  Example Simulation Results

Based on the model described in the previous section, a number of simulations have been performed. Some example simulation traces are included in this section as an illustration; see Figures 4 and Figure 5 (here the time delays within the temporal LEADSTO relations were taken 1 time unit). In Figure 4 two different traces are shown with different characteristics. Note that the scaling of the vertical axis differs per graph. For both traces the world state shows an offer with a rather modest strength of $0.3$. Moreover both $\gamma_1 = 0.6$ and $\gamma_2 = 0.6$. Simulation trace 1 at the left hand side has $\beta_1 = 0.5$ and $\beta_2 = 0.5$, whereas simulation trace 2 at the right hand side has $\beta_1 = 0.5$ and $\beta_2 = 1$. In trace 1 the belief (and also the feeling) gets the same strength as the stimulus, namely $0.3$; here no effect of the emotional response is observed. However, in trace 2 the belief gets a higher strength (namely $0.65$) due to the stronger emotional response (with feeling getting strength $1$). This shows how a belief can be affected in a substantial manner by the feedback from the emotional response on the belief.

In Figure 5 the complete example scenario for the car parking case discussed earlier is shown. The world state shows something that (from a distance) looks like a charge with strenght $0.8$ until time point 225; this is indicated by the dark line in the upper part of Figure 5. For this case $\beta_1 = 0.8$ and $\beta_2 = 0.4$ was taken, which means a modest role for the emotional response. The belief in a charge leads to an increasingly strong emotional body state $b1$ and via the related feeling, the belief reaches a strength a bit above $0.9$.

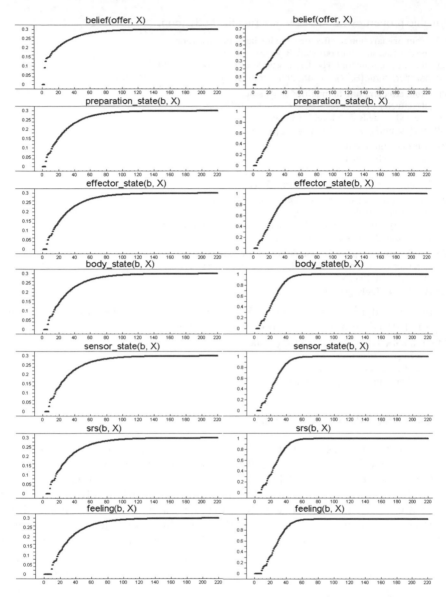

**Fig. 4.** Two example traces: (1) $\beta_1 = 0.5$ and $\beta_2 = 0.5$, (2) $\beta_1 = 0.5$ and $\beta_2 = 1$

However, having come closer to the car, after time point 225 the world state shows with strength *0.8* something that is more like an offer, whereas the strength of the charge shown drops to *0.05*, which also was the strength of the offer before time point 225. As a consequence the belief in a charge drops and based on a different emotion response on the offer belief based on body state *b2* the strength of the belief in an offer increases until above *0.9*.

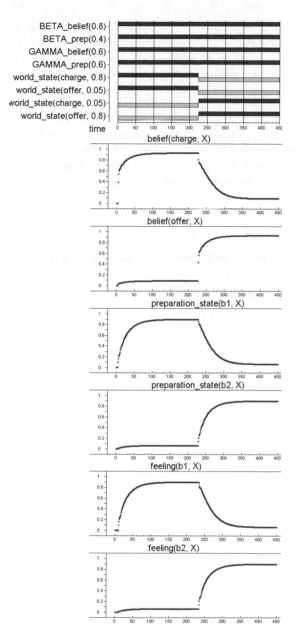

**Fig. 5.** Trace for the car parking case with $\beta_1 = 0.8$ and $\beta_2 = 0.4$

## 5 Mathematical Analysis

In the example simulations discussed above it was shown that for a time period with a constant environment, the strengths of beliefs, body states and feelings reach a stable equilibrium. By a mathematical analysis it can be addressed which types of equilibria are possible. To this end equations for equilibria can be determined from the dynamical model equations for the belief and the preparation state level, which can be expressed as differential equations as follows (with $b(t)$ the level of the belief, $s(t)$ of the stimulus, $f(t)$ of the feeling, and $p(t)$ of the preparation for the body state at time $t$).

$$db(t)/dt = \gamma_1 (\beta_1(1-(1-s(t))(1-f(t))) + (1-\beta_1)s(t)f(t) - b(t))$$
$$dp(t)/dt = \gamma_2 (\beta_2(1-(1-b(t))(1-f(t))) + (1-\beta_2)b(t)f(t) - p(t))$$

To obtain equations for equilibria, constant values for all variables are assumed (also the ones that are used as inputs such as the stimuli). Then in all of the equations the reference to time $t$ can be left out, and in addition the derivatives $db(t)/dt$ and $dp(t)/dt$ can be replaced by $0$. Assuming $\gamma_1$ and $\gamma_2$ nonzero, this leads to the following equations.

$$\beta_1(1-(1-s)(1-f)) + (1-\beta_1)sf - b = 0$$
$$\beta_2(1-(1-b)(1-f)) + (1-\beta_2)bf - p = 0$$

As for an equilibrium it also holds that $f = p$, this results in the following two equations in $b$, $f$, and $s$:

$$\beta_1(1-(1-s)(1-f)) + (1-\beta_1)sf - b = 0 \qquad (1)$$
$$\beta_2(1-(1-b)(1-f)) + (1-\beta_2)bf - f = 0 \qquad (2)$$

For the general case (1) can directly be used to express $b$ in $f$, $s$ and $\beta_1$. Using this, in (2) $b$ can be replaced by this expression in $f$, $s$ and $\beta_1$, which transforms (2) into a quadratic equation in $f$ with coefficients in terms of $s$ and the parameters $\beta_1$ and $\beta_2$. Solving this quadratic equation algebraically provides a complex expression for $f$ in terms of $s$, $\beta_1$ and $\beta_2$. Using this, by (1) also an expression for $b$ in terms of $s$, $\beta_1$ and $\beta_2$ can be found. As these expressions become rather complex, only an overview for a number of special cases is shown in Table 1 (for 9 combinations of values $0$, $0.5$ and $1$ for both $\beta_1$ and $\beta_2$). For these cases the equations (1) and (2) can be substantially simplified as shown in the second column (for equation (1)) and second row (for equation (2)). The shaded cases are instable (not attracting), so they only occur when these values are taken as initial values.

As can be seen in this table, for persons that are pessimistic for believing ($\beta_1 = 0$) and have a negative profile in generating emotional responses ($\beta_2 = 0$), reach a stable equilibrium for which both the belief and the feeling have level $0$. The opposite case occurs when a person is optimistic for believing ($\beta_1 = 1$) and has a positive profile in generating emotional responses ($\beta_2 = 1$). Such a person reaches a stable equilibrium for which both the belief and the feeling have level $1$. For cases where one of these $\beta_1$ and $\beta_2$ is $0$ and the other one is $1$, a stable equilibrium is reached where the belief gets the same level as the stimulus: $b = s$. When a person is in the middle between optimistic and pessimistic for believing ($\beta_1 = 0.5$), for the case of a negative profile in generating emotional responses the stable belief reached gets half of the level of the stimulus, whereas for the case of a positive profile in generating emotional responses the stable belief reached gets $0.5$ above half of the level of the stimulus (which is the $0.65$ shown in the second trace in Figure 4). This clearly shows the effect of the feeling on the belief. The case where both $\beta_1 = 0.5$ and $\beta_2 = 0.5$ is illustrated in the first trace in Figure 4: $b = f = s$.

**Table 1.** Overview of equilibria for 9 cases of parameter settings

| $\beta_2$ | | 0 | | 0.5 | 1 | |
| $\beta_1$ | eq. (1) | eq. (2)    $f = 0$ | $b = 1$ | $b = f$ | $f = 1$ | $b = 0$ |
|---|---|---|---|---|---|---|
| $0$ | $b = sf$ | $b = f = 0$ | $b = f = s = 1$ | $b = f = 0$<br>$b = f$ and $s = 1$ | $b = s$<br>$f = 1$ | $b = s = 0$<br>$b = f = 0$ |
| $0.5$ | $b = (s + f)/2$ | $b = s/2$<br>$f = 0$ | $b = f = s = 1$ | $b = f = s$ | $b = (s + 1)/2$<br>$f = 1$ | $b = f = s = 0$ |
| $1$ | $1-b = (1-s)(1-f)$ | $b = s$<br>$f = 0$ | $b = f = 1$<br>$b = s = 1$ | $b = f = 1$<br>$b = f$ and $s = 0$ | $b = f = 1$ | $b = f = s = 0$ |

# 6  Discussion

In this paper an agent model was introduced incorporating the reciprocal interaction between believing and feeling based on neurological theories that address the role of emotions and feelings. A belief usually triggers an emotional response. Conversely, a belief may not only depend on information obtained, but also on this emotional response, as, for example, shown in literature such as (Eich et al., 2000; Forgas et al., 2005; Forgas et al., 2009; Niedenthal, 2007; Schooler and Eich, 2000). In the literature, this phenomenon has been studied informally but no formal computational models have been developed, as far as the authors know. Accordingly, this paper is an attempt to develop a formal computational model of how a belief generates an emotional response that is felt, and on the other hand how the emotion that is felt affects the belief. For feeling the emotion, based on elements taken from (Damasio, 1999, 2004; Bosse, Jonker and Treur, 2008), a converging recursive body loop is included in the model. As a second loop the model includes a converging feedback loop for the interaction between feeling and belief. The causal relation from feeling to belief in this second loop was inspired by the Somatic Marker Hypothesis described in (Damasio, 1994, 1996; Bechara and Damasio, 2004), and may also be justified by a Hebbian learning principle (cf. Hebb, 1949; Bi and Poo, 2001), as also has been done for the functioning of mirror neurons; e.g., (Keysers and Perrett, 2004; Keysers and Gazzola, 2009). Both the strength of the belief and of the feeling emerge as a result of the dynamic pattern generated by the combination of the two loops. The model was specified in the hybrid dynamic modelling language LEADSTO, and simulations were performed in its software environment; cf. (Bosse, Jonker, Meij, and Treur, 2007). A mathematical analysis of the equilibria of the model was discussed. The model was illustrated using an example scenario where beliefs are affected by negative and positive emotional responses.

# References

1. Bechara, A., Damasio, A.: The Somatic Marker Hypothesis: a neural theory of economic decision. Games and Economic Behavior 52, 336–372 (2004)
2. Bi, G.Q., Poo, M.M.: Synaptic Modifications by Correlated Activity: Hebb's Postulate Revisited. Ann. Rev. Neurosci. 24, 139–166 (2001)

3.  Bosse, T., Jonker, C.M., van der Meij, L., Treur, J.: A Language and Environment for Analysis of Dynamics by Simulation. International Journal of Artificial Intelligence Tools 16, 435–464 (2007)
4.  Bosse, T., Jonker, C.M., Treur, J.: Simulation and Analysis of Adaptive Agents: an Integrative Modelling Approach. Advances in Complex Systems Journal 10, 335–357 (2007)
5.  Bosse, T., Jonker, C.M., Treur, J.: Formalisation of Damasio's Theory of Emotion, Feeling and Core Consciousness. Consciousness and Cognition Journal 17, 94–113 (2008)
6.  Bosse, T., Jonker, C.M., Los, S.A., van der Torre, L., Treur, J.: Formal Analysis of Trace Conditioning. Cognitive Systems Research J. 8, 36–47 (2007)
7.  Damasio, A.: Descartes' Error: Emotion, Reason and the Human Brain. Papermac, London (1994)
8.  Damasio, A.: The Somatic Marker Hypothesis and the Possible Functions of the Prefrontal Cortex. In: Philosophical Transactions of the Royal Society: Biological Sciences, vol. 351, pp. 1413–1420 (1996)
9.  Damasio, A.: The Feeling of What Happens. In: Body and Emotion in the Making of Consciousness. Harcourt Brace, New York (1999)
10. Damasio, A.: Looking for Spinoza. Vintage books, London (2004)
11. Eich, E., Kihlstrom, J.F., Bower, G.H., Forgas, J.P., Niedenthal, P.M.: Cognition and Emotion. Oxford University Press, New York (2000)
12. Forgas, J.P., Laham, S.M., Vargas, P.T.: Mood effects on eyewitness memory: Affective influences on susceptibility to misinformation. Journal of Experimental Social Psychology 41, 574–588 (2005)
13. Forgas, J.P., Goldenberg, L., Unkelbach, C.: Can bad weather improve your memory? An unobtrusive field study of natural mood effects on real-life memory. Journal of Experimental Social Psychology 45, 254–257 (2009)
14. Hebb, D.: The Organisation of Behavior. Wiley, New York (1949)
15. Jonker, C.M., Snoep, J.L., Treur, J., Westerhoff, H.V., Wijngaards, W.C.A.: BDI-Modelling of Complex Intracellular Dynamics. Journal of Theoretical Biology 251, 1–23 (2008)
16. Keysers, C., Gazzola, V.: Unifying Social Cognition. In: Pineda, J.A. (ed.) Mirror Neuron Systems: the Role of Mirroring Processes in Social Cognition, pp. 3–28. Humana Press/Springer Science (2009)
17. Keysers, C., Perrett, D.I.: Demystifying social cognition: a Hebbian perspective. Trends in Cognitive Sciences 8, 501–507 (2004)
18. Niedenthal, P.M.: Embodying Emotion. Science 316, 1002–1005 (2007)
19. Schooler, J.W., Eich, E.: Memory for Emotional Events. In: Tulving, E., Craik, F.I.M. (eds.) The Oxford Handbook of Memory, pp. 379–394. Oxford University Press, Oxford (2000)
20. Winkielman, P., Niedenthal, P.M., Oberman, L.M.: Embodied Perspective on Emotion-Cognition Interactions. In: Pineda, J.A. (ed.) Mirror Neuron Systems: the Role of Mirroring Processes in Social Cognition, pp. 235–257. Humana Press/Springer Science (2009)

# Information Hypothesis: On Human Information Capability Study

Jiří Krajíček

Faculty of Information Technology, Brno University of Technology, Department of
Information Systems, Božetěchova 2, Brno 612 66, Czech Republic
ikrajice@fit.vutbr.cz

**Abstract.** Main aim of this paper is to explore the human information capabilities with link to open problems in computer science. We come with working hypothesis reflecting currently known research experimental evidence of human information capabilities. As every hypothesis, presented hypothesis needs further verification to show confirmation or disconfirmation in result. Nevertheless, this work opens novel topic on scientific research with the aim to resolve presented open problems and review of classical paradigm in computer science.

## 1 Introduction

The computing industry has passed parallel hardware revolution with the introduction of more-core processors (more than one computational core) and parallel software revolution is right incoming. Beside proposed parallel challenge in hardware and software design [3], even toady we can observe certain limitations we are facing [22].

In effort to find solutions for open problems (see critical review for details) here we turn research focus back to nature approach, to human base. To show relevant scientific contributions and publications, we can highlight Lucas and Penrose contributions, for example. Lucas, in his paper *"Minds, machines and Gödel"* [20], is arguing that the human mathematician cannot be represented by any machines and Penrose [29] has suggested that the human mind might be the consequence of quantum-mechanically enhanced, "*non-algorithmic*" computation. Penrose uses variation of the "*halting problem*" to show that mind cannot be an algorithmic process [30]. Rosen has proposed that computation is an inaccurate representation of natural causes that are in place in nature [33]. Moreover Kampis [16] assumes that information content of an algorithmic process is fixed and no "*new*" information is brought forward. These publications are a few examples of different information capability (to compute and communicate) between human and classical computer and represent motivation for novel research approach. Hence here we are interested in researching open problems in inner scope rather than outer (from human to human, through human). Presented information hypothesis is an effort to approximate human information capability and highlight its advantages.

In second chapter we introduce critical review from point of open problems, walls in computer science. In third chapter main alternative approaches as related work are briefly described. The chapter fourth introduces another alternative approach – human information capability and continues by proposed information hypothesis on human

N. Zhong et al. (Eds.): BI 2009, LNAI 5819, pp. 25–35, 2009.

information model, which is the research synthesis of scientific evidence related to human information process. The fifth chapter briefly discusses future work and preparing experiments which will be essential for future research and hypothesis verification.

## 2 Critical Review: Open Problems in Computer Science

### 2.1 Physical Walls

Computer science has already faced physical walls in implementation of logical gates, in silicon chip (concerning size, overheating, unsustainable power consumption). Therefore researchers and chip industries are switching to HW and SW paralleliza- tion. Nowadays parallel switch help us to survive in trend of performance growth (approximately doubling every 18 months) [3]. But once it is done (maximal parallel speedup is reached) there is no obvious clue how to continue this settled performance growth (one cannot parallelize task that was completely parallelized before) [6].

Here we can also recall Feynman [8] research question: *"How small can you make a computer?"* Gordon Moor himself is also expecting that his law is limited by eco- nomics, including the rising cost of advanced lithography and new wafer fabs [22]. Although advanced lithography is still promising the decreasing (shrinking) of chip size design, at current rate of lithography scaling (~0.7 every two years), the industry will reach the length scale of atomic crystal lattices (order 1nm) by mid 2040s.

### 2.2 Theoretical Walls

Theoretical walls are open questions closely related to computational model (Turing machine, TM) and its implementation consequences which are restricted by physical limitation (as described above). Here we can list undecidable problems like (Halting Problem, Post Correspondence Problem), NP-hard and NP-complete problems like (Travelling Salesman Problem, SAT Problem, Knapsack Problem, Graph Coloring Problem). However, any problem in category of undecidable problems (e.g. Halting problem) or in category of NP-complete walls (e.g. 3-SAT problem) is not efficiently solvable by using classical computation methodology.

Beside the fact there exist theoretical computational models more powerful than TM, like TM with Oracle, Iterative TM with advice, Site machines (reflecting capa- bility of distributed computing and modern personal computer non-uniform computa- tion), real number models, variants of neural networks with real numbers [34], due to the infinite descriptive properties it is not possible to implement it efficiently and use in practice. Thus effective *"artificial"* computable theoretical model remains bounded by Turing machine.

In spite the Turing Machine model is formally described, closely related Church- Turing hypothesis is still an open question itself and needs formalization so that it can be finally proved or disproved (confirmed or disconfirmed).

### 2.3 Human-Computer Walls

Beside the physical and theoretical walls computer science is also dealing with many human-computer interaction (HCI) problems/issues which are consequences of classical computation design and development [25]. From functionality principle computer, as machine, is diametrically different from human. Therefore people are educated to understand computers, to think like computers. It is simply observable that as greater difference exist at interaction between human and computer as more efforts must be spent to eliminate human-computer gap (gulf of execution, gulf of evaluation) [24].

We can heavy change the human computable design, but we can decide how "*machine*" computation is designed. Therefore related science of cognitive psychology and modeling is positively contributive by studying and testing brain functions. In presented hypothesis we understand human functions as special cases of "*computation and communication*" not strictly limited to physical brain activity (see chapter fifth for details).

### 2.4  Discussion: Questions Identified

We have briefly listed main open problems of computer science. In future researchers would like to handle physical, theoretical and human-computer walls as presented. Therefore these walls represent the main motivation for alternative computation. By considering alternative computing approaches we may push forward several research questions.

- Does the computation have to be processed exactly by the way of classical computation?
- What is necessary to change in computer science (research paradigm) to not face these walls/problems, what is possible to do to resolve these problems?
- Is there any way of computation/communication not restricted to these theoretical, physical, HCI walls?
- Beside the artificial TM, is there any natural model which is more powerful and more efficient?
- If there is any, what are the main limitations/walls of such approach?

## 3  Related Work: Alternative Computation

As mentioned in previous chapter, open problems, walls are the main motivation for alternative computable approaches. In searching responses for questions as arisen from faced walls (see critical review above) and to propose novel inspiration in alternative natural computation [17], let us cite from [28]:

*"The universe doesn't need to calculate, it just does it. We can take the computational stance, and view many physical, chemical and biological processes as if they were computations: the Principle of Least Action "computes" the shortest path for light and bodies in free fall; water "computes" its own level; evolution "computes" fitter organisms; DNA and morphogenesis "computes" phenotypes; the immune system "computes" antigen recognition. This natural computation can be more effective than a digital simulation. Gravitational stellar clusters do not "slow down" if more stars are added, despite the problem appearing to us to be $O(n^2)$. And as*

*Feynman noted, the real world performs quantum mechanical computations exponentially faster than can classical simulations".*

As we can see, real world computing is breaking the Turing paradigm [28]. In following we briefly summarize two main representatives of real world alternative computing (quantum, DNA computation) and discuss its properties.

Both, quantum and DNA are not restricted to silicon chip HW platform (dealing physical walls), come from nature, real world observation, both are massive parallel (dealing theoretical walls) and promising to be more efficient.

## 3.1  Quantum Computation

There is still an open question whether all quantum mechanical events are Turing-computable, although it is known that classical TM is the same powerful as quantum TM and vice versa [7]. However Deutsch [7], Bernstein and Vazirani [4] shoved that there exist problems that quantum TM solve problems faster than classical one, can be more efficient.

For instance, Grover's quantum searching algorithm compute with $N$ entries in $O(N1/2)$ time and using $O(\log N)$ storage space [9]. In contrast to classically, searching an unsorted database requires a linear search, which is $O(N)$ in time. Also Shor [27] demonstrated that a quantum computer could efficiently solve a well-known problem of integer factorization, i.e. finding the factors of a given integer $N$, in a time which is polynomial in $\log N$. This is exponentially faster than the best-known classical factoring algorithm. Quantum computing was also successfully used to "*treat*" (finding better "*solutions*" than classical computation can reach) some of artificial intelligence NP-hard problems [23, 15].

Although quantum computing is promising to overcome physical walls of classical silicon chip design and decrease time complexity at case of certain theoretical problems, it cannot be used in practice due to its „*own*" walls as seems today. Main disadvantages are:

- Expensive to fabricate.
- Today quantum 16-qubit chip (D-Wave) is not enough.
- Decoherence, unwanted interaction with environment during the computation, the lost of superposition state.

## 3.2  DNA Computation

Operations of molecular genetics can be understood as operations over strings/strands and by proper sequence of such operations we are able to execute required computation. In 1994 Adleman [1], inventor of DNA computation, demonstrated how to use DNA to solve a well-known NP-complete problem, the "*travelling salesman*" problem in polynomial time. The goal of the problem is to find the shortest route between numbers of cities, going through each city only once. Adleman chose to find the shortest route between 7 cities. In later years DNA computation was generalized to solve other NP-complete problems and other specific problems like SAT problem, 3-SAT problem [5]. Although the DNA computing is also promising to overcome physical walls of classical silicon chip design and solve NP-complete problems efficiently, the reality is different due to space complexity expansion. Briefly main disadvantages are:

- Representation of problem leads to non optimal space complexity.
- Execution time (e.g. one computational step can take one day or more).
- Risk of errors during biological operations (requires repeating until acceptable).

### 3.3  Discussion

Alternative computable approaches like Quantum or DNA computation come with "*partial*" solution, novel approach, but still have other "*own*" problems. We may ask another research questions.

- What is the next step?
- Is there any other alternative approach which can positively contribute to open problems/walls?

## 4  The Study: Human Information Capabilities

As shown in previous chapter alternative approaches are candidates to overcome walls in computer science, on the other hand still not feasible, applicable. In this paper we consider another alternative approach – human information capabilities, which stand for human computation and communication abilities. As human information capabilities are still undiscovered research area (the power and time complexity of human information processing is still unknown) we focus on this topic to contribute positive on presented walls.

Classic computation (as we know today) provides two main services to human for satisfying his needs. These closely related services are:

- Processing information, form of information may differ (dealing with computation).
- Sharing information (dealing with communication).

Both of these services can be found on human in natural form. Further, there exist specific tasks "*solvable*" by human computation, which cannot be solved by any classical computer or which is hard to solve (e.g. artificial intelligence NP-hard problems) [23, 2]. Here we can note some specific tasks like pattern recognition, path navigation, natural language processing or more abstract tasks like creativity, free will and being consciousness. Moreover, beyond these well-known tasks there is also experimental evidence on human information capabilities which cannot be classically explained on basis of established physical concepts and statistical theory [13, 26, 19].

In summary, we do not know the diversity, power and complexity of human information capabilities and moreover we do not know it's all causers and principles [10, 12, 36]. For instance, there is experimental evidence which links mind process with holonomic quantum interpretation [31], but quantum interpretation is still matter of open question itself.

It is not possible to answer such open questions in one paper; hence here we operate with information which is known, classical scientific concepts (e.g. neural networks) and results of experimental evidence (e.g. unexplainable human phenomena) to propose research synthesis – working hypothesis with aim to contribute on open problems, walls in computer science.

## 4.1 Research Objectives

Here we briefly list main research objectives as concludes from research motivation and proposed working hypothesis (see section 4.2 below).

- Proposal of working hypothesis on human information capabilities motivated by open problems in computer science (this paper).
- To approximate model of human information capabilities related to working hypothesis (briefly this paper).
- Proposal of experiments: to verify presented information hypothesis, to design computable/communication patterns and to contribute on walls in computer science (briefly this paper).
- Perform of proposed experiments (future work).

In conclusion to explore human information capabilities and explain how human computation may be performed via designed human computable model.

## 4.2 Working Information Hypothesis

In classical computer science there exist many human/natural based approaches (e.g. iterative evolutionary approach). These approaches are commonly studied separately in contrast to real world evidence. Here we assume these approaches as cooperative system, research synthesis rather than stand alone approach (to be closer to real human). This also reflects real evidence of human scientific results (psychology, neurobiology, and quantum mechanics) [12, 36, 30, 10]. Whole hypothesis is divided into three parts and linked together. Now, we postulate hypothesis statements consequently.

### 4.2.1 Working Hypothesis ($H_1$)
*Rather than describing human information capability like independent neural network, iterative evolutionary computation or fuzzy system, etc, we are assuming the synthesis of these approaches as cooperative information (computing and communicating) system based on neural networks, molecular neurobiology, evolutionary approach and phenomena related to quantum mechanics at least (as proposed on reference information human model, see figure 4.1).*

In real world human mind, thinking is related to brain functions, where brain includes biological neural network (here is the reference to level of neural network computation).

Biological neural network describes a population of interconnected neurons or a group of disparate neurons whose inputs or signaling targets define a recognizable circuit. This circuit is evolvable and reconfigurable [21] (here is the link for iterative evolvable approach).

The field of molecular neurobiology overlaps with other areas of biology and chemistry, particularly genetics and biochemistry and study behaviors on molecular level, behind neuron size. Study of nervous behavior on such level brings us to microscopic world, molecular level where quantum phenomena can be examined. Further the evidence on such level of non-local physical, chemical and biological effects supports quantum brain theory [12, 36]. These theories where proposed by various researchers [30, 31, 32, 10] (here is the link which supports quantum phenomena explanation).

Although clear quantum explanation on this level remains open question until verification, at least we now consider the research evidence of activity which is beyond the neuron size (see below).

### 4.2.2  Working Hypothesis ($H_2$)

*By considering and supporting all phenomena in cooperative information system as stated in $H_1$ information process in human information model (see figure 4.1) is positively changed in terms of complexity/efficiency, in contrast with considering each phenomenon as independent model, stand alone information processing approach.*

With respect to $H_1$ and $H_2$ statement, recently in computer science it has been proven that assuming "*ingredients*" of real modern computing like non-uniformity of programs, interaction of machines and infinity operations in cooperative model, simultaneously; leads to model beyond Turing machine, to the model of Site machine, Turing machine with advice [34], see also chapter 2.2.

### 4.2.3  Working Hypothesis ($H_3$)

*Based on $H_1$ and $H_2$ computation and communication patterns can be designed to replace or overcome some of problems/tasks we are attempting to solve in computer science.*

With respect to $H_1$, $H_2$ and $H_3$ statement; further research in following third statement can also bring us explanation why human is able to solve NP-hard problems similarly to quantum computation; meanwhile these problems are not efficiently solvable by classical computation today [23, 2].

### 4.3  Towards Human Information Model

Meanwhile human science is referring to psychology (natural SW), biology, medicine (natural HW); the technical science is referring to classical computer science and physics. Here we have to meet both research attitudes to fulfill common goal. To verify the presented hypothesis statements $H_1$, $H_2$ and $H_3$ we need to deal with human information model. In general, we may try to describe the best model for human information capabilities, but the best model will always remain human himself. Moreover it is not currently feasible to answer all human-information open questions in one model. Therefore presented model stands for links of "*what do we know*" separately to observe "*what we need to overcome*". At this phase, we do not need to answer all questions to be able to use what we have, what we are.

Figure 1 present general multi-level model of human information capabilities. The existence of each level is examined further and was confirmed by experimental evidence as described above.

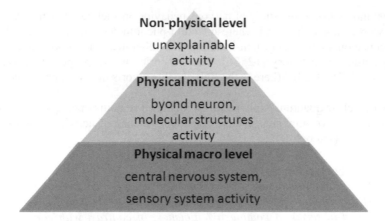

**Fig. 1.** General multi-level model of human information capabilities, each level is linked with others as cooperative system. This is a reference model, pointer to human capabilities.

### 4.3.1  Physical Macro Level

Physical macro level refers to nervous system linked with sensory preceptors which can be described as neural network connected to external devices. It is still an open question what variant of neural network is the best approximation for human-natural network. Currently the model of Analogue Recurrent Neural Networks (ARNN) is examined as human neural network candidate. If ARNN operates with real numbers then computable power is equal to Turing machine with oracle [35]. However the evolutionary gene extension, biological growth and die aspects in neurons seem to be essential for biological neural network approximation, modeling [21].

### 4.3.2  Physical Micro Level

Physical micro level refers to evidence of computation beyond the neuron level. Molecular neurobiology has discovered that biochemical transactions play an important role in neuronal computations; on level where toady quantum mechanics is the most accurate description. For instance, in [32] dendrite spine is examined as a quantum computing device. Further, recently the Shafir-Tversky statistical effect (well known in cognitive psychology) was represented by quantum model [18]. Moreover there is huge experimental evidence on phenomena (beyond neuron level) which supports quantum explanation [12, 36, 31, 11].

### 4.3.3  Non-physical Level

Non-physical refers to human information activity which cannot be explained on basis of established physical concepts and statistical theory. It is assumed that such activity is executed beyond any physical artifact/part of human body [26, 13, 14]. Although the explanation of e.g. paranormal scientific evidence [13], near death experiences [19] is matter of discussion and open questions, we should consider such evidence as part of human information capabilities as evidence shows.

According to conventional scientific view the mind is product of the brain, neural network activity (electro-chemical reactions between neurons). This is based on observation of metabolic brain activity in response to specific thinking process. But in critical examination this observation only implies the role of such neurons as mediator and does not necessary imply the origin of thought itself. The studies based on NDE (near death experiences) and paranormal evidence might be useful in support of this information level which is assumed to be beyond any physical artifact (brain) and might put the neural network just to role of mediator, receiver rather than origin (the fundamental information level).

## 5 Experimental Verification: Proposal of Experiment Tasks

In future work, we are going to focus more on experiments, where human can execute specific task (generally interested in tasks which are easy for human but difficult for computers). In these experiments human activity (on each modeled level) can be recorded (e.g. by EEG, fMRI). In more detail, we focus on study of connections between synaptic functions in neuronal level and quantum coherence inside neuron especially among microtubules (micro-level) [10, 11].

Each level provides certain resources which are essential in task solving. Here we are interested in what impact has each phenomenon on each level, what activity is recorded and what resources of recorded activity are available.

We have selected several essential experiments which need to be tested. Designed experiments include testing of following tasks (the most of presented tasks naturally come from human - artificial intelligence):

- Pattern recognition (CAPTCHA application [2] testing, e.g. morphed letter recognition).
- Image comparison (NP-complete problem: two images of the same place from different observer view are compared).
- Image description (Open computer problem: describe images by finite sequence of words to identify it).
- Sentence development (human will develop such nature sentence that is an open problem for any classical computer).
- Further investigation in near death experience testing (interested in human information processing beyond physical brain EEG activity, EEG totally flat). For examples of research evidence see [19].

As we can see, the proper execution (performing) of proposed experiments naturally requires further research work. Here we have briefly described experimental task's form and properties to link between this and future work, between presented hypotheses (theory) and hypothesis verification (analysis of experimental evidence).

## 6 Conclusion

We have listed main open problems of computer science, discussed alternative approaches which are promising the solutions. But in spite of quantum, DNA alternative efforts open problems/walls remains open. Hence we focused on human information capabilities as other alternative approach and presented the working information hypothesis which is offering the positive contribution on discussed problems. We have also presented the abstract human information model which is essential for future investigation in hypothesis verification and preparing experiments.

Beside these proposed experiments, as main research objective remains the verification of presented hypothesis, mostly based on human experimental evidence, human model approximation and final design of patterns ("*algorithmic-rules*") for problems which we are able to solve in terms of $H_3$ statement.

# References

1. Adleman, L.: Molecular Computation of Solutions to Combinatorial Problems. Science 266, 1021–1024 (1994)
2. Ahn, L., et al.: CAPTCHA: Using Hard AI Problems for Security. In: Biham, E. (ed.) EUROCRYPT 2003. LNCS, vol. 2656, pp. 294–311. Springer, Heidelberg (2003)
3. Asanovic, K., Bodik, R., et al.: The Landscape of Parallel Computing Re-search: A View from Berkeley, white paper (2006)
4. Bernstein, E., Vazirani, U.: Quantum complexity theory. In: Proc. 25th Annual ACM Symposium on Theory of Computing, pp. 11–20. ACM, New York (1993)
5. Braich, et al: Solution of a 20-variable 3-SAT problem on a DNA computer. Scienceexpress (2002)
6. Leiserson, C.E., Mirman, I.B.: How to Survive the Multicore Software Revolution, Cilk Arts, white paper (2008)
7. Deutsch, D.: Quantum theory, the Church-Turing principle and the universal quantum computer. Proceedings of the Royal Society of Lon-don A400 (1985)
8. Feyman, R.P.: Feyman Lectures on Computation. In: Hey, A., Allen, R. (eds.) Penguin Books, pp. 182–183 (1996)
9. Grover, L.K.: A fast quantum mechanical algorithm for database search. In: Proceedings of the 28th Annual ACM Symposium on the Theory of Computing, p. 212 (1996)
10. Hameroff, S.: Funda-Mentality: is the conscious mind subtly linked to a basic level of the Universe? Trends Cognitive Science 2, 4–6 (1998)
11. Hameroff, S.: The Brain Is Both Neurocomputer and Quantum Computer. Cognitive Science Society 31, 1035–1045 (2007)
12. Hu, H., Wu, M.: Nonlocal effects of chemical substances on the brain produced through quantum entanglement. Progress in Physics 3, 20–26 (2006)
13. Jahn, R.B.: On the Quantum Mechanics of Consciousness, with Application to Anomalous Phenomena. Foundations of Physics 16, 721–772 (1986)
14. Jung, C.G.: Psychic conflicts in a child. In: Collected Works of C. G. Jung, vol. 17. Princeton University Press, Princeton (1970)
15. Kaminsky, W.M., Lloyd, S.: Scalable Architecture for Adiabatic Quantum Computing of NP-Hard. In: Quantum Computing & Quantum Bits in Mesoscopic Systems. Kluwer Academic, Dordrecht (2003)
16. Kampis, G.: Self-Modifying systems in biology and cognitive science. Pergamon Press, New York (1991)
17. Krajicek, J.: A Note on Application of Natural Phenomena in Computer Science. In: Proceedings of the 14th Conference EEICT 2008, vol. 4, pp. 1–5 (2008)
18. Khrennikov, A.: Quantum-like model of cognitive decision making and information processing. Biosystems 95(3), 179–187 (2009)

19. Lommel, P., et al.: Near-death experience in survivors of cardiac arrest: a prospective study in the Netherlands. The Lancet (2001)
20. Lucas, J.R.: Minds, machines and Gödel. In: Anderson, A.R. (ed.), Minds and Machines, pp. 43–59. Prentice-Hall, Englewood Cliffs (1954)
21. Miller, J., et al.: An Evolutionary System using Development and Artificial Genetic Regulatory Networks. In: 9th IEEE Congress on Evolutionary Computation (CEC 2008), pp. 815–822 (2008)
22. Moor, G.: Lithography and the future of Moor's Law. In: Proceeding SPIE, vol. 2439, pp. 2–17 (1995)
23. Neven, H., et al.: Image recognition with an adiabatic quantum computer I. Mapping to quadratic unconstrained binary optimization, eprint arXiv:0804.4457 (2008)
24. Norman, D.: The Design of Everyday Things. Doubleday Business (1990)
25. Qiyang, C.: Human Computer Interaction: Issues and Challenges. Idea Group Publishing (2001)
26. Sheldrake, R.: An experimental test of the hypothesis of formative causation. Rivista di Biologia – Biology Forum 86, 431–444 (1992)
27. Shor, P.: Polynomial-Time Algorithms for Prime Factorization and Discrete Logarithms on a Quantum Computer. In: Proceedings of the 35th Annual Symposium on Foundations of Computer Science, pp. 20–22 (1994)
28. Stepney, S.: Jourenyes in non-classical computation, I: A grand challenge for computing research. The international Journal of Parallel, Emergent and Distributive Systems 20(1), 1–9 (2005)
29. Penrose, R.: The Emperor's New Mind: Concerning Computers, Minds, and The Laws of Physics. Oxford University Press, Oxford (1989)
30. Penrose, R.: Shadows of the Mind: A search for the missing science of consciousness. Oxford University Press, Oxford (1994)
31. Pribram, K.H.: Rethinking Neural networks: Quantum Fields and Bio-logical Data. In: Proceedings of the first Appalachian Conference on Behavioral Neurodynamics. Lawrence Erlbaum Associates Publishers Hillsdale, Mahwah (1993)
32. Rocha, A.F., et al.: Can The Human Brain Do Quantum Computing? Medical Hypotheses 63(5), 895–899 (2004)
33. Rosen, R.: Life Itself: A comprehensive Inquiry into the nature, origin, and fabrication of life. Columbia University Press, New York (1991)
34. Wiedermann, J.: The Turing Machine Paradigm in Contemporary Computing. In: Mathematics Unlimited - 2001 and Beyond, pp. 1139–1155. Springer, Heidelberg (2000)
35. Zenil, H., et al.: On the possible computational power of the human mind. In: Essays on Epistemology, Evolution, and Emergence. World Scientific, Singapore (2006)
36. Wu, M., Hu, M.: Evidence of non-local physical, chemical and biological effects supports quantum brain. NeuroQuantology 4(4), 291–306 (2006)

# Correlated Size Variations Measured in Human Visual Cortex V1/V2/V3 with Functional MRI

Tianyi Yan[1], Fengzhe Jin[3], and Jinglong Wu[1,2]

[1] Graduate School of Natural Science and Technology, Okayama University
3-1-1 Tsushima-naka, Okayama, Japan
[2] International WIC Institute, Beijing University of
Technology, Beijing, China
wu@mech.okayama-u.ac.jp
[3] Design Department of Transportation Systems Works Transportation
Division Toyo Denki Seizo K.K., Japan

**Abstract.** The retinotopic characteristics on human peripheral vision are still not well known. The position, surface area and visual field representation of human visual areas V1, V2 and V3 were measured using fMRI in 8 subjects (16 hemispheres). Cortical visual field maps of the 120deg were measured using rotating wedge and expanding ring stimuli. The boundaries between areas were identified using an automated procedure to fit an atlas of the expected visual field map to the data. All position and surface area measurements were made along the boundary between white matter and gray matter. In this study, we developed a new visual presentation system widest view (60 deg of eccentricity). The wide-view visual presentation system was made from nonmagnetic optical fibers and a contact lens, so can use in general clinical fMRI condition and the cost is lower. We used the newly wide view visual presentation system, the representation of the visual field in areas V1, V2 and V3 spans about 2223 mm^2,1679 mm^2 and 1690 mm^2 .

## 1 Introduction

The number of neurons in human visual cortex far exceeds the number in many other species that depend on vision. For example, the surface area of macaque monkey visual cortex is probably no more than 20% that of human visual cortex, although the cortical neuronal density is similar. This species difference in the number of cells in visual cortex cannot be explained by how these species encode the visual world. The monkey samples the retinal image at a higher resolution. There are 1.5 million optic nerve fibers from each eye in macaque and only 1 million such fibers in humans [1]. The large size of human visual cortex is likely not a result of an increase in the supply of information but, rather, due to an increase in visual processing and the organization and delivery of information to other parts of cortex, such as those devoted to language and reading. Given these differences in visual cortex size, it would not be surprising that many features of human visual cortex are not present in closely related primate systems [2].

N. Zhong et al. (Eds.): BI 2009, LNAI 5819, pp. 36–44, 2009.
© Springer-Verlag Berlin Heidelberg 2009

In human visual pathway, visual information passes from retina to lateral genicu-late nucleus and then to human primary visual cortex. Therefore, human primary visual cortex (V1) contains a map of visual space. To a good approximation, each two-dimensional (2D) location in the visual field is represented at a single physical location within V1. Area V1 is the human visual cortical area with the most well-defined anatomical boundaries, agreed on by virtually all previous studies, both his-torical and more recent [3, 4].

Previous retinotopic studies of humans by fMRI have identified the V1 [5], placing it between the occipital pole and the lateral end of the parieto-occipital sulcus (POS) [6]. Andrews et al. (1997) measured the size of the lateral geniculate nucleus (LGN) and the optic tract as well as the surface area of striate cortex. They observed that the surface area correlates closely with the cross-sectional area of the optic tract as well as with the area and volume of the lateral geniculate nucleus (LGN). Given that pho-toreceptor density also varies by up to a factor of three across individuals [7], it is possible that this density is a key variable that leads to the variation in size of the central representations found in the LGN and V1. To what extent do the sizes of other visual areas follow the size of V1? This question has not been answered precisely. Amunts et al. (2000) measured the volume of Brodmann's areas 17 and 18 in ten brains (post-mortem). However, they did not report on a correlation between the sizes of these areas. Also, the correspondence between Brodmann's area 18 and visual area V2 is not as clear as that between striate cortex and V1 [8].

In this study, we investigated the quantitative relationship between the V1 surface area and human peripheral visual field and estimated the cortical surface area of the V1, V2, V3. We used a newly developed visual presentation system to examine the V1, V1, V2, V3 surface area and areal cortical magnification for 0° to 60° of eccentricity. Therefore human V1 appeared to be an ideal location to test for addition functional features within a well defined, well accepted cortical area, by using functional MRI (fMRI).

## 2   Method and Materials

### 2.1   Subjects and Stimuli

Eight healthy subjects without previous neurological or psychiatric disorders (age 19–31 years, mean 25 years; two women, six men) participated in the study. The subjects had normal or corrected-to-normal vision and were right-handed. Visual stimuli were created on a display using a resolution of 800 × 600 pixels. The display stimulus was brought to the subject's eyes within the scanner by a wide-view optical-fiber presenta-tion system. Monocular (right eye) presentations were accomplished; the optical-fiber screen (surface-curved with a curvature radius of 30 mm) was placed in the center of the 30 mm from a subject's eye. The visual field of stimulus was a 120° horizontal × 120° vertical. Because the screen so was close to the eye, subjects wore a contact lens (Menicon soft MA; Menicon, Japan. with +20, +22, +25 magnification) to retain their length of focus. We obtained written informed consent from all subjects before the experiment. The study was approved by the Institutional Research Review Board of Kagawa University, Japan.

To identify the retinotopic areas of the visual cortex, we carried out fMRI scans while subjects viewed phase-encoding stimuli [8]. A high-contrast, black-and-white, radial checkerboard pattern (mean luminance 110cd/m^2, contrast 97%) reversed contrast at a frequency of 8 Hz [9], with eccentricity ranging from a 0° to 60° visual angles. Two types of stimulus were used for locating visual area boundaries and estimating eccentricity. The stimulus for locating boundaries was a 22.5° wedge that rotated slowly counterclockwise about a red fixation spot at the center of the stimuli (As Fig.1 Polar). The wedge rotated in steps of 22.5°, remaining in each position for 4 s before instantaneously rotating to the next position. The stimulus for estimating eccentricity was an expanding checkered annulus. The flickering radial checkerboard was moved from the center to the periphery in discrete steps (each step 7.5°, with a total of eight steps, As Fig.1 Eccentricity), remaining at each position for 8s before instantaneously expanding to the next position.

## 2.2  MR Data Acquisition

The fMRI experiment was performed using a 1.5 T Philips clinical scanner (Intera Achieva; Best, The Netherlands). All images were acquired using a standard radio-frequency head coil. We acquired 23 slices approximately orthogonal to the calcarine sulcus to cover most of the cortical visual areas. The T2*-weighted gradient echo-planner imaging sequence was used with the following parameters: TR/TE = 2000/50 ms; FA = 90°; matrix size = 64 × 64; and voxel size = 3 × 3 × 3 mm. Before acquiring the functional images, T2-weighted anatomical images were obtained in the same planes as the functional images, using the spin echo sequence. A T1-weighted high-resolution image was also acquired after each functional experiment.

## 2.3  Data Analysis

The functional and anatomical data were processed using BrainVoyager software package (Brain Innovation, Masstricht, Netherlands). After preprocessing the functional data, anatomical data was processed. The recorded high-resolution T1-weighted three-dimensional (3-D) recordings were used for surface reconstruction. The gray and white matter was segmented using a region-growing method, and the white matter cortical surface was reconstructed. Prior to surface flattening, the cortical surface was inflated and cut along the calcarine from the occipital pole to slightly anterior of the POS [10].

The functional data was aligned onto the 3-D anatomic image using the image co-ordinates. To identify boundaries (wedge stimuli), maps were created based on cross-correlation values for each voxel, determined by a standard hemodynamic box-car function ($r \geq 0.25$). We identified the boundaries of the V1 by hand based on the horizontal and vertical meridians and knowledge of the retinotopic organization of the visual cortex.

All volume measurements were made on the 3-D anatomical image. Under each eccentricity condition (0°–7.5°, 7.5°–15°, 15°–22.5°, 22.5°–30°, 30°–37.5°, 37.5°–45°, 45°–52.5°, 52.5°–60°), each strongest-response voxel in the V1 was counted as active for each 3-D anatomical image. Because the voxel size is 1 × 1× 1 mm, the active voxel volume in the V1 for each eccentricity condition equals the counted number of voxels. If we assume that cortical thickness was invariable in the V1, then the V1 surface area can be obtained by dividing the voxel volume by the cortical thickness, assumed here to be 2.5 mm^2.

# 3  Results

## 3.1  Eccentricity Maps

Figure 1 shows a three-dimensional rendering of the left hemisphere of subject LS. The surface represents the boundary between white and gray matter that was identified by segmentation algorithm [11]. Within this portion of the brain, each cortical region responds mainly to a visual stimulus at one retinotopic location.

Figure 1A shows a color map of the response to an expanding ring on a medial view of the cortical surface, which indicates the eccentricity (distance from the fovea) that causes a signal at that location. The hue of the color at each cortical surface point indicates the response phase, which is proportional to the eccentricity of the local visual field representation. The data in this figure represent the average of three separate sessions. In Figure 1B, the cortical surface was unfolded, which is processed by inflation algorithm. In Figure 1C, the surface region including the occipital lobe, posterior parts of the parietal lobe and temporal lobe, containing the activated area has been cut off, and the resulting approximately conical surface cut again along the fundus of the calcarine sulcus to allow it to be flattened completely.

There is a systematic increase in eccentricity moving anteriorly along the medial wall of occipital cortex. As the expanding ring stimulus moved from the fovea to the periphery of the retina, the location of the responding areas varied from posterior to anterior portions of the calcarine sulcus in what is referred to as the eccentricity dimension of retinotopic. The larger peripheral representation crossed to the fundus of the parieto-occipital sulcus.

A parallel treatment of data from the rotating hemifield stimulus is shown in Figure1. The color also indicates the phase of the periodic response, which is now proportional to the polar angle of the local visual field representation. The locations of several visual areas were identified by measuring angular visual field representations [12, 13]. Figure 1d shows the angular visual field representations on the folded and unfolded surface that spans most of the occipital lobe and posterior parts of the parietal lobe and temporal lobe. The color at each location represents the angle of the rotating wedge that caused an fMRI response.

## 3.2  The Sizes of V1/V2/V3

Table 1 contains measurements of the surface area of the visual field representations from 0-60 deg. Right and left hemispheres, (left and right visual field) as well as dorsal and ventral aspects (lower and upper visual field) are listed separately for each subject. Figure 2 shows the correlation in size between V2 and V3 is significant

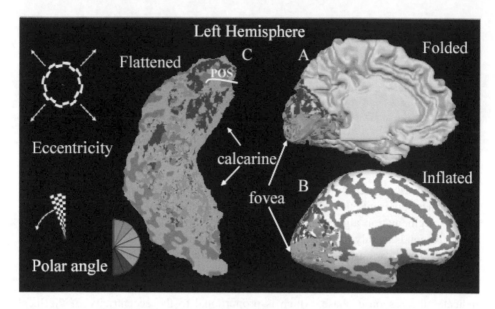

**Fig. 1.** Eccentricity and Polar angle maps of human visual areas. The top row shows polar angle coded by color  displayed on the folded cortical in the left hemifield head (A), the cut and inflated cortical surface (B), and the flattened cortical (C).

**Table 1.** Surface area Measurements

| Subject | Hemi. | Ventral | | | Dorsal | | | Total | | |
|---------|-------|------|------|------|------|------|------|------|------|------|
| | | V1 | V2 | V3 | V1 | V2 | V3 | V1 | V2 | V3 |
| BY | left | 1207 | 1115 | 897 | 895 | 1013 | 878 | 2102 | 2012 | 1775 |
| BY | right | 1019 | 965 | 756 | 1209 | 998 | 762 | 2228 | 1721 | 1518 |
| DB | left | 986 | 578 | 576 | 1388 | 729 | 674 | 2374 | 1154 | 1250 |
| DB | right | 883 | 921 | 839 | 1033 | 927 | 631 | 1916 | 1760 | 1470 |
| FZ | left | 1187 | 776 | 774 | 1033 | 884 | 799 | 2220 | 1550 | 1573 |
| FZ | right | 1013 | 924 | 730 | 1147 | 950 | 760 | 2160 | 1654 | 1490 |
| HO | left | 1179 | 1131 | 1025 | 1231 | 1182 | 870 | 2410 | 2156 | 1895 |
| HO | right | 1147 | 1639 | 1115 | 1071 | 987 | 870 | 2218 | 2754 | 1985 |
| KS | left | 746 | 957 | 886 | 1076 | 790 | 610 | 1822 | 1843 | 1496 |
| KS | right | 1060 | 1126 | 886 | 1104 | 908 | 809 | 2164 | 2012 | 1695 |
| LS | left | 951 | 980 | 699 | 950 | 810 | 681 | 1901 | 1679 | 1380 |
| LS | right | 1044 | 1212 | 792 | 984 | 1017 | 799 | 2028 | 2004 | 1591 |
| NK | left | 1425 | 805 | 477 | 1170 | 529 | 599 | 2595 | 1282 | 1076 |
| NK | right | 1047 | 908 | 664 | 1220 | 990 | 781 | 2267 | 1572 | 1445 |
| WH | left | 895 | 872 | 891 | 1472 | 805 | 589 | 2367 | 1763 | 1480 |
| WH | right | 1155 | 1088 | 919 | 1319 | 844 | 863 | 2474 | 2007 | 1782 |
| | left mean | 1072 | 901 | 778 | 1151 | 842 | 712 | 2223 | 1679 | 1490 |
| | right mean | 1046 | 1097 | 837 | 1135 | 952 | 784 | 2181 | 1935 | 1622 |
| | total mean | 1059 | 999 | 807 | 1143 | 897 | 748 | 2202 | 1807 | 1556 |
| | stdev | 159 | 232 | 160 | 158 | 149 | 103 | 215 | 370 | 232 |

Table 1. Surface area measurements for the 60 deg visual field representation of V1, V2 and V3. The measurements are shown for right and left hemispheres, dorsal and ventral aspects of V1/2/3, and eight different subjects (16 hemispheres). Various summary statistics are listed at the bottom of the table.

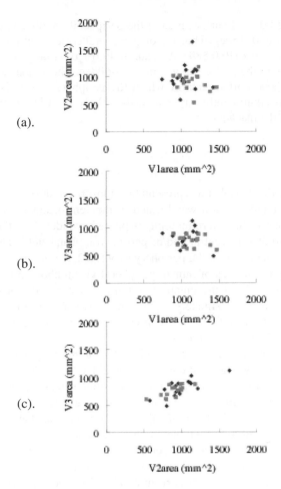

(a).

(b).

(c).

**Fig. 2.** V2 surface area correlates with V3 surface area (c), but V1 and V2 (a), V1 and V3 (b) surface area is no significant correlation. Note that these are measurements of quarter-field cortical representations. Triangles are ventral regions and squares are dorsal regions data.

(r = 0.761, p< 0.001, df = 32), but between V1 and V2 (r = 0.072, p=0.696, df = 32), V1 and V3 (r = 0.231, p=0.204, df = 32) were not.

The V2 surface area in the wide-filed 60 deg representation is roughly 82% that of V1, and this size difference is statistically significant (pairwise t = 3.249, p < 0.01, df = 31). The surface area of V3 is averaged 1556mm^2. V3 is on average 71% the size of V1, and this size difference is statistically significant (pairwise t = 7.790, p < 0.001, df = 31).

We measured the average areas of V1/V2/V3 in different degree as a function of eccentricity, as shown figure 2. There was a significant decrease of the response area in V1(y = -123.98Ln(x) + 553.99, R1=0.9822), V2(y = -82.69Ln(x) + 372.58, R2=0.9898) and V3(y = -75.329Ln(x) + 313.91, R3=0.9876). The tread of decreasing of V2/V3 is slower than that of V1 and the areas of V1/V2/V3 in 50deg. The trend of decreasing slowed down in 40~50deg of V2/V3, which is different with V1.

We also measured the V1 surface area of the peripheral 15~60deg and compared it with that of the central 0~15deg. The area on peripheral cortex is almost the same size with that on central cortex (P=0.889). A mean of 49.6% of striate cortex was devoted to the representation of the peripheral vision in 15-60deg. The data is not significantly different from the post-mortem data of which the peripheral cortex is about 46.9%. The percentage was significantly larger than the 39.1% of striate cortex devoted to peripheral vision in the macaque.

## 4 Discussion

Our results indicate that peripheral representation most reach dorsorostral by the POS and most subjects' representation was located in the ventrocaudal bank of the POS. Previous fMRI studies of human retinotopic mapping have identified the V1 using the visual stimulus of limited to central and/or peri-central visual field. The stimuli used did not directly activate much of the periphery in V1, V2, V3 and it was hard to compare with physiological studies of human peripheral vision above 30° of eccentricity [14]. Using wide-view visual presentation system, we estimated that the V1 has an average surface size of approximately 2223 mm^2, which represents the portion of visual field eccentricity from 0° to 60°. Our estimates are consistent with the results of those physiological and fMRI studies.

Were the cortical surface a plane, one could calculate the gray matter volume from the surface area and knowledge of the mean cortical thickness. The cortical surface is not flat, and in regions of high curvature the local volume can differ measurably from the estimate based on planarity. Specifically, volume is underestimated on the crowns of gyri and overestimated in the fundi of sulci. For large regions that include sulci and gyri, these two errors tend to cancel one another. For smaller regions these two types of errors may not cancel well. In a subset of the regions reported in this paper, we calculated the difference between estimates of gray matter volume assuming planarity and estimates that account for local curvature. The difference between the two estimates never exceeded 5 percent, even in small regions such as V3 or the sub-regions used to estimate cortical magnification. Hence, for the regions we report here, it is reasonable to estimate the gray matter tissue volume as surface area multiplied by cortical thickness.

Does surface area correlate with performance? Duncan and Boynton (2002) have reported a correlation between cortical magnification estimates (based on surface area) and a visual acuity task [15]. If such a correlation is observed in several contexts, then a theory relating the size of the neuronal substrate, say based on signal-to-noise ratio, may become accepted. Should this connection become secure, then the analysis of correlation between surface area and performance may provide a means for uncovering the functional role of visual areas.

V1 surface area (mean = 2202 mm^2) is larger than previous research which focus on central vision [16]. The area mapped out to 60 deg is about twice of which in central vision. The area for V1 surface agrees with the post-mortem data.

The previous research found a correlation between the surface area of V1 and the size of the retinal and geniculate input streams. And, Dougherty et al. made the investigation that the surface areas of V1 and V2, V2 and V3 have a relatively high correlation, between V1 and V3 was not. In our results, the correlation in size between V2 and V3 is significant, but between V1 and V2, V1 and V3were not. The V2 surface area in the wide-filed 60 deg representation is roughly 82% that of V1, and this size difference is statistically significant. The surface area of V3 is averaged 1556 mm^2. V3 is on average 71% the size of V1, and this size difference is statistically significant. There is a difference from previous research in central vision [17]. In Dougherty's data, the surface area of V2 was about 75% and V3 was about 55% the size of V1. The area of V2 and V3 in our results were significatively larger than that in 0~12deg. In our results, the correlation in size between V2 and V3 is significant, but between V1 and V2, V1 and V3 were not. The previous DTI research [18] found the pattern of connections between V1/V2 ↔ PPA and V3 ↔ FFA, respectively. So among V1, V2 and V3 may be complex correlation because of the influence of additional factors, such as the insertion of a significant pulvinar input at the level of V2, V3 and increasing significance of feedback and other projections.

## 5   Conclusion

In this study, we quantitatively investigated the functional characteristics of human peripheral visual fields. We used fMRI to identify the areal magnification factor and surface area size of the V1, V2, V3 for a wide visual field of eccentricity up to 60° which is lager than that in previous studies. We also estimated the average V1, V2, V3 surface area. This study demonstrated that among V1, V2 and V3 may be complex correlation because of the influence of additional factors, such as the insertion of a significant pulvinar input at the level of V2, V3 and increasing significance of feedback and other projections.

## Acknowledgment

A part of this study was financially supported by JSPS AA Science Platform Program and Grant-in-Aid for Scientific Research (B) (21404002). We thank the subjects who participated in this study and the staff of the Osaka Neurosurgery Hospital for their assistance with data collection.

## References

1. Engel, S.A., Glover, G.H., Wandell, B.A.: Retinotopic organization in human visual cortex and the spatial precision of functional MRI. Cereb. Cortex 7, 181–192 (1997)
2. Sereno, M.I., Dale, A.M., Reppas, J.B., Kwong, K.K., Belliveau, J.W., Brady, T.J., et al.: Borders of multiple visual areas in humans revealed by functional magnetic resonance imaging. Science 268, 889–893 (1995)
3. DeYoe, E.A., Carman, G.J., Bandettini, P., Glickman, S., Wieser, J., Cox, R., et al.: Mapping striate and extrastriate visual areas in human cerebral cortex. In: Proc. Natl. Acad. Sci. USA, vol. 93, pp. 2382–2386 (1996)
4. Duncan, R.O., Boynton, G.M.: Cortical magnification within human primary visual cortex correlates with acuity thresholds. Neuron 38, 659–671 (2003)

5. Van Essen, D.C., Newsome, W.T., Maunsell, J.H.: The visual field representation in striate cortex of the macaque monkey: asymmetries, anisotropies, and individual variability. Vision Res. 24, 429–448 (1984)
6. Adams, D.L., Horton, J.C.: A precise retinotopic map of primate striate cortex generated from the representation of angioscotomas. J. Neurosci. 23, 3771–3789 (2003)
7. Andrews, T.J., Halpern, S.D., Purves, D.: Correlated size variations in human visual cortex, lateral geniculate nucleus, and optic tract. J. Neurosci. 17, 2859–2868 (1997)
8. Stensaas, S.S., Eddington, D.K., Dobelle, W.H.: The topography and variability of the primary visual cortex in man. J. Neurosurg. 40, 747–755 (1974)
9. Horton, J.C., Hoyt, W.F.: The representation of the visual field in human striate cortex. In: A revision of the classic Holmes map. Arch Ophthalmol, vol. 109, pp. 816–824 (1991)
10. Pitzalis, S., Galletti, C., Huang, R.S., Patria, F., Committeri, G., Galati, G., et al.: Widefield retinotopy defines human cortical visual area v6. J. Neurosci. 26, 7962–7973 (2006)
11. Stenbacka, L., Vanni, S.: Central luminance flicker can activate peripheral retinotopic representation. Neuroimag. 34, 342–348 (2007)
12. Dougherty, R.F., Koch, V.M., Brewer, A.A., Fischer, B., Modersitzki, J., Wandell, B.A.: Visual field representations and locations of visual areas V1/2/3 in human visual cortex. J. Vis. 3, 586–598 (2003)
13. Qiu, A., Rosenau, B.J., Greenberg, A.S., Hurdal, M.K., Barta, P., Yantis, S., et al.: Estimating linear cortical magnification in human primary visual cortex via dynamic programming. Neuroimage 31, 125–138 (2006)
14. Ejima, Y., Takahashi, S., Yamamoto, H., Fukunaga, M., Tanaka, C., Ebisu, T., et al.: Interindividual and interspecies variations of the extrastriate visual cortex. Neuroreport 14, 1579–1583 (2003)
15. Smith, A.T., Singh, K.D., Williams, A.L., Greenlee, M.W.: Estimating receptive field size from fMRI data in human striate and extrastriate visual cortex. Cereb. Cortex 11, 1182–1190 (2001)
16. Goebel, R., Khorram-Sefat, D., Muckli, L., Hacker, H., Singer, W.: The constructive nature of vision: direct evidence from functional magnetic resonance imaging studies of apparent motion and motion imagery. Eur. J. Neurosci. 10, 1563–1573 (1998)
17. Lu, H., Basso, G., Serences, J.T., Yantis, S., Golay, X., van Zijl, P.C.: Retinotopic mapping in the human visual cortex using vascular space occupancy-dependent functional magnetic resonance imaging. Neuroreport 16, 1635–1640 (2005)
18. Fischl, B., Dale, A.M.: Measuring the thickness of the human cerebral cortex from magnetic resonance images. In: Proc. Natl. Acad. Sci. USA, vol. 97, pp. 11050–11055 (2000)

# Effects of Attention on Dynamic Emotional Expressions Processing

Liang Zhang[1,2], Brigitte Roeder[3], and Kan Zhang[1]

[1] Institute of Psychology, Chinese Academy of Sciences,
100101, Beijing, China
[2] Graduate University of Chinese Academy of Sciences,
100049, Beijing, China
[3] Biological Psychology and Neuropsychology,
20146, Hamburg, Germany
zhangl@psych.ac.cn

**Abstract.** Attention and emotion both play the crucial roles in human cognitive processing. This study tried to investigate the relationship between attention and emotion, using dynamic facial expressions which are natural and frequently encountered in everyday life. The results showed that the emotional expressions are processed faster than the neutral ones when they are outside the current focus of attention. It indicates that the emotion processing is automatic and not gated by the attention.

## 1 Introduction

Attention and emotion both play the crucial roles in human cognitive processing. Emotion is one of the basic survival-related factors. It produces specific bodily responses, aimed at preparing the organism for crucial behavior. Specialized neural systems are evolved for the rapid perceptual analysis of emotionally salient events, such as emotional facial expressions. (M. Eimer, A. Holmes, 2003).

Meanwhile, numerous stimuli from the environment confront our limited processing capacity simultaneously. The attention mechanism helps the brain select and process only those stimuli most relevant to the ongoing behavior. But adaptive behavior requires to monitor the environment and detect potential survival-related stimuli (e.g., emotional) even when they are unexpected and not current task relevant or are outside the focus of attention [2].

The notion that attention bias to the emotion was supported by some studies. They provide the evidence that detection of emotional stimuli occurs rapidly and automatically [3]. A controversial issue is whether the encoding and analysis of emotionally salient events can occur independently of attention [4]. Recent studies provided conflicting results. In an fMRI study [5], spatial attention was manipulated by having subjects respond to stimulus arrays containing two faces and two non-face stimuli (houses). Four stimuli were presented with a cross in the center. Faces presented to the left and right of the cross, and houses presented below and above the cross. Or vice versus. In each trial, subjects either compared the two faces or the two houses. Thus, the attention was manipulated on the face pair or on the house pair. And facial

N. Zhong et al. (Eds.): BI 2009, LNAI 5819, pp. 45–52, 2009.
© Springer-Verlag Berlin Heidelberg 2009

expression was either fearful or neutral. The results showed that the fMRI response to fearful and neutral faces was not modulated by the focus of attention, consistent with the view that the processing of emotional items does not require attention.

However, the opposite results came up later in another ERP study by Holmes and Vuilleumier [1]. When faces were attended, a greater frontal positivity in response to arrays containing fearful faces than the neutral faces. In contrast, when faces were unattended, this emotional expression effect was completely eliminated. This study demonstrates a strong attentional gating in the emotion processing.

The ERP result was supported by an fMRI research by Pessoa [6]. Participants were instructed to focus on the gender of faces or on the orientation of the bars. A Face in the centre and two bars at peripheral were presented simultaneously. The participants' task was either to judge the gender of the face or to determine whether the two bars had the same orientation. Thus the spatial attention was controlled on or off the face. The bar-orientation task was made very difficult to consume most attentional resources, leaving little resource to process the unattended faces. During the gender task, fearful faces evoked stronger neural activity than neutral faces in a network of brain regions including the fusiform gyrus, superior temporal sulcus, orbitofrontal cortex and amygdala. Note that such different activation was only observed in the gender task but not observed in the bar-orientation task.

Another study by Anderson[7]investigated this question by manipulating object-based attention while keeping spatial attention constant. Double-exposure' images that contained faces and buildings were used. Both of them were semi-transparent, and subjects were instructed to make either a male/female judgment (attend to faces) or an inside/outside judgment (attend to places). No effect of attention was observed for the two expressions. Similar responses were evoked to both attended and unattended fearful or neutral faces in the amygdala. However, an interesting effect was observed with expressions of disgust, which evoked stronger signals in the amygdala during unattended relative to attended conditions [8].

With only a few exceptions [9, 10, 11], studies on perception emotional expressions were conducted using static faces as stimuli. However, neuroimaging studies have revealed that the brain regions known to be implicated in the processing of facial affect, such as the posterior superior temporal sulcus (pSTS), the amygdala and the insula, respond more to dynamic than to static emotional expressions [12]. So it is more appropriate, in research dealing with the recognition of real-life facial expressions, to use dynamic stimuli [11].

The present study thus attempts to investigate the relationship between attention and emotion. To do this we use a standardized set of dynamic facial expressions with angry, happy and neutral emotions, which are presented either attended or unattended as a result of manipulation of spatial attention.

## 2   Methods

### 2.1   Participants

Fifteen volunteers participated in the experiment. All the participants had normal or corrected-to-normal vision. The participants received course credits or were paid award for taking part in the study.

## 2.2  Materials

Videos that consisted of faces of four different individuals (two male, two female) were used. The videos consist of face articulating a nonsense bi-syllable word (without the audio track). Each of the stimuli was expressed by a kind of emotion valence: happy, angry or neutral.

Mean duration of standard (non-target) stimuli is 1170 ms. There is no significant difference among the three different emotional types of stimuli. Deviant stimuli were inserted with a 200 ms interruption on the basic of the standard stimuli. All faces covered a visual angle of about 3.4*4.9 degree.

## 2.3  Design and Procedure

This is a two-factor within-subject design. The experimental factors are spatial attention (attended/unattended) and emotion valence (angry/happy/ neutral).

Participants were seated in a dimly lit sound-attenuated room, with response buttons under their right hands. All stimuli were presented on a computer screen in front of a black background at a viewing distance of 70cm. A cross maintained at the center of the screen as the fixation. The eccentricity of the faces (measured as the distance between the centre of each face and the central fixation cross) was 4.2 degree.

The experiments consist of six blocks, each containing 48 standard trials and 16 deviant trials. In each block, participants were asked to pay attention to only one side (left or right) and respond to the deviants and standards as soon as possible. Forty-two standards and six deviants were presented at the attended side in every block. In each trial, an expression video was present at only one side of the screen with the central fixation. The ISI varied from 1200ms to 1500ms.

# 3   Results

One participant's data are excluded due to the low accuracy (below 80%). Fourteen available data were collected and analyzed. Considering the deviants with short interruption were artificially made and can not present the natural expressions, the responds to the deviant stimuli are excluded. Only the data of the standard stimuli are analyzed and compared through different conditions. Repeated Measure ANOVAs with two factors, facial emotion (angry, happy, neutral) and spatial attention were carried out on percentage of accurate responses and reaction time, respectively.

## 3.1  Reaction Time

The reaction time from the onset of the stimulus to the respond was recorded. The average of RT to all the stimuli is 1214.76ms.

The main effect of spatial attention is significant ($F_{1, 13} = 12.506$, $P<0.01$). The responds to the stimuli in the attended location is much faster than the responds to the unattended stimuli.

RTs of emotional and neutral expressions are different. If we consider all the data regardless the attention factor, the reaction to the emotional stimuli is faster than the neutral stimuli but not significant ($F_{2, 26} = 2.623$, $P=0.092$). But Fig.1 shows the detail

**Table 1.** RTs of the different expressions under attended/unattended condition (ms)

|         | Attended          | Unattended        |
|---------|-------------------|-------------------|
| Angry   | 1212.23 ± 59.87   | 1235.97 ± 55.10   |
| Happy   | 1212.04 ± 37.95   | 1223.02 ± 55.62   |
| Neutral | 1220.02 ± 65.51   | 1257.81 ± 60.85   |
| Mean    | 1214.76 ± 54.45   | 1238.94 ± 57.19   |

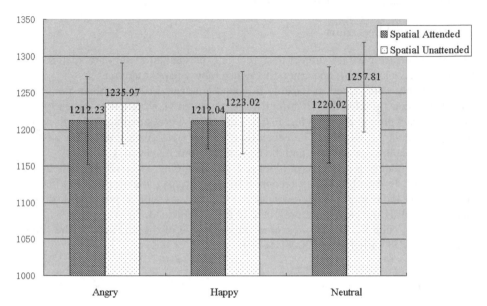

**Fig. 1.** RTs of the expressions with different emotion valences under attend/unattended condition

**Table 2.** Paried T-test on RTs of expressions with different emotion valences under attended/unattended condition (ms)

| Mean Differences |         | Spatial Attended | | | Spatial Unattended | | |
|------------------|---------|-------|-------|---------|-------|-------|---------|
|                  |         | Angry | Happy | Neutral | Angry | Happy | Neutral |
| Spatial Attended | Angry   |       |       |         | 23.74 |       |         |
|                  | Happy   | 0.18  |       |         |       | 10.98 |         |
|                  | Neutral | -7.79 | -7.97 |         |       |       | 37.80** |
| Spatial Unattended | Angry |       |       |         |       |       |         |
|                  | Happy   |       |       |         | 12.95 |       |         |
|                  | Neutral |       |       |         | -21.84* | -34.79* |       |

between spatial attention and emotions. There is no difference between emotional and neutral ones when the expressions were presented at the attended location. But there is an obvious difference among emotional and neutral expressions when the stimuli were presented at the unattended location.

Further analysis by Pairwise Comparisons showed that RTs to the unattended angry expression is higher than unattended neutral expression (Diff. = 21.844ms, marginal significant $P=0.053$). So does RTs of happy expressions (Diff. = 34.792 ms, $P < 0.05$).

Paried T-test was carried out on the RTs of angry, happy, and neutral expressions, respectively. Results showed that the reaction to the emotional expressions varied indistinctively with the attention, but distinctively to the neutral expression ($t_{1, 13} = 3.474$, $P<0.01$). RTs of neutral expressions became much slower when the stimuli showed up on the unattended side. But for angry and happy expressions, the reaction changes from attended to unattended location were not significant.

## 3.2 Accuracy

The overall performance was high. (Mean$_{Accuracy}$ = 82.60%). The accuracy among all the conditions was above 80% except the responds to the unattended neutral expression (79.76%). See Figure 2.

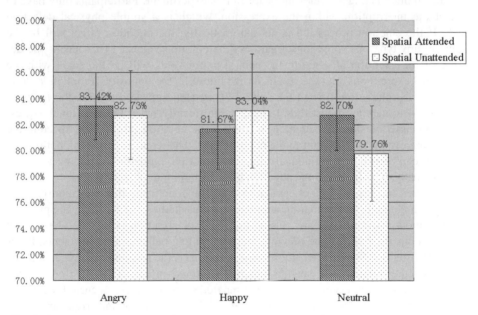

**Fig. 2.** The accuracies of responds to the expressions with different emotion valences under attend/unattended conditions

Neither the main effect of emotion nor spatial attention is significant at the aspect of accuracy (Emotion $F_{2, 26}$ = .420; Attention $F_{1, 13}$ = .292). Also, there is no interaction between the emotion and attention ($F_{2, 26}$ = .478).

The accuracies of the emotional expressions are the same under the spatial attended condition and the unattended condition. There is a slight difference between the responds to neutral expression under the two attended conditions but didn't reach the significance.

## 4 Discussion

The primary aim of the present experiment was to investigate the relationship between the attention and emotion. We use dynamic facial expressions as the emotional stimuli, which are natural and frequently encountered in everyday life. The spatial attention was manipulated by instructing the participants to pay attention to one side of the screen thus maintain the attended focus at the certain location. The probability of the stimuli is another complement to control the spatial attention. In each block, 75% stimuli were present at the attended side. A sustained attention paradigm was employed (with left/right side attended tasks delivered in separate experimental blocks).

It is predictable that the process will be slower when the stimuli appear outside the current focus of attention. This is confirmed by the gain of RTs to the stimuli at the attended location comparing with unattended location. The question is whether the emotional expressions are processed pre-attention or automatically.

The emotion is a task-irrelevant factor in this experiment. Participants only have to detect the interruption within the expression, which is a simple physical detection task. If the emotion processing is not automatically, the attention effect will be the same to any type of expressions regardless the emotion valence. That means the decline of the attention absence to the neutral expressions will be the same with the emotional expressions. But the results turned out to be the opposite. It took more time to distinguish the stimuli when the neutral expression appeared outside the focus of attention. But increased cost didn't change significantly when the emotional expressions appeared outside the attention. It implies the emotion information speedup the processing even it is irrelevant to the ongoing task. The result supports the notion that emotion can be processed automatically. The emotion expression drew the attention automatically, which proved the attention bias to emotional objects again.

However, our results are not enough to draw the conclusion that the emotion processing is pre-attention. Comparing with the previous studies, evidence for the processing of emotion stimuli that are outside the focus of attention is mixed. The present result is consistent with some previous study including behavioral experiments, ERPs and fMRI studies. Many fMRI studies pay attention on the respond of amygdala to the fear, and other related area, such as superior temporal sulcus (STS), fusiform cortex [5]. A left amygdala response to fearful compared to neutral faces occurred regardless of whether the faces were at relevant/attended locations or at the irrelevant/ unattended locations, demonstrating that fear processing in the amygdala was obligatory and unaffected by the modulation of spatial attention.

The mixed conclusion can be related to the spare processing capacity that is utilized for the processing of task-irrelevant or unattended items. The studies that revealed that attention modulates the processing of emotional stimuli employed very demanding tasks that might have nearly exhausted the processing capacity. By contrast, the studies that observed little or no effect of attention used less demanding tasks. The task in this experiment is relative simple and with little processing resource involved as our design. It was proved by the high accuracy and the participants' feedback. Plenty spare processing capacity can be utilized during the task. It is a reason to explain the fast processed to the unattended emotional items.

Besides, the difference only existed between the emotional and neutral expressions, and we didn't find the difference between the positive emotion (happy) and negative emotion (angry). Although there were some studies found the process of positive emotion and negative emotion activated different brain areas and may involve different processing mechanism [13, 6]. Emotional valence should not be neglected when we discuss the attention effect on the emotion processing. In the study by Anderson [7], the amygdala responses to attended and unattended fearful faces were the same, responses to unattended disgusted faces were, paradoxically, increased. Unfortunately, the two negative emotions were not in our experiment. The further research is going to consider more emotion valence.

In summary, the study with dynamic natural expressions found the emotional effect under the unattended condition. It indicates that the expressions with emotional information are processed faster than the ones without emotion (e.g., neutral) when they are outside the current focus of attention. As the salient information for the survival and the keys in human social communication, the emotion requires fast processed. The processing to the emotion is automatic regardless it is inside or outside the attention and regardless it is relevant or irrelevant to the ongoing behavior. Further research should utilize electrophysiology or neuroimaging to test the behavioral results. It will also reveal the details of emotion processing mechanism and the role of attention plays.

# References

1. Eimer, M., Holmes, A., McGlone, F.: The role of spatial attention in the processing of facial expression: An ERP study of rapid brain responses to six basic emotions. Cognitive, Affective, and Behavioral Neurosci. 3, 97–110 (2003)
2. Vuilleumier, P., Armony, J.: Reciprocal links between Emotion and Attention Human. In: Brain Function, 2nd edn., pp. 419–444. Academic Press, San Diego (2003)
3. Globisch, J., Hamm, A.O., Esteves, F., Ohman, A.: Fear appears fast: temporal course of startle reflex potentiation in animal fearful subjects. Psychophysiology 36, 66–75 (1999)
4. Holmes, A., Kiss, M., Eimer, M.: Attention modulates the processing of emotional expression triggered by foveal faces. Neurosci. Lett. 394, 48–52 (2006)
5. Vuilleumier, P., Armony, J.L., Driver, J., Dolan, R.J.: Effects of Attention and Emotion on Face Processing in the Human Brain: An Event-Related fMRI Study. Neuron 30, 829–841 (2001)
6. Pessoa, L., McKenna, M., Gutierrz, E., Ungerleider, L.G.: Neural processing of emotional faces requires attention. In: Proc. Natl. Acad. Sci. USA, vol. 99, pp. 11458–11463 (2002)
7. Anderson, A.K., Christoff, K., Panitz, D., De Rosa, E., Gabrieli, J.D.: Neural correlates of the automatic processing of threat facial signals. J. Neurosci. 23, 5627–5633 (2003)
8. Pessoa, L.: To what extent are emotional visual stimuli processed without attention and awareness? Current Opinion in Neurobiology 15, 188–196 (2005)
9. de Gelder, B., Bocker, K.B., Tuomainen, J., Hensen, M., Vroomen, J.: The combined perception of emotion fromvoice and face:early interaction revealed by human electric brain responses. Neurosci. Lett. 260, 133–136 (1999)

10. Kreifelts, B., Ethofer, T., Grodd, W., Erb, M., Wildgruber, D.: Audiovisual integration of emotional signals in voice and face: an event-related fMRI study. Neuroimage 37, 1445–1456 (2007)
11. Collignon, O., Girard, S., Gosselin, F., Roy, S., Saint-Amour, D., Lassonde, M., Lepore, F.: Audio-visual integration of emotion expression. Brain Research 1242, 126–135 (2008)
12. Kilts, C.D., Egan, G., Gideon, D.A., Ely, T.D., Hoffman, J.M.: Dissociable neural pathways are involved in the recognition of emotion in static and dynamic facial expressions. Neuroimage 18, 156–168 (2003)
13. Armony, J.L., Dolan, R.J.: Modulation of spatial attention by fear-conditioned stimuli: an event-related fMRI study. Neuropsychologia 40, 817–826 (2002)

# Simulating Human Heuristic Problem Solving: A Study by Combining ACT-R and fMRI Brain Image

Rifeng Wang[1,2], Jie Xiang[3,1], Haiyan Zhou[1], Yulin Qin[1,4], and Ning Zhong[1,5]

[1] The International WIC Institute, Beijing University of Technology, China
[2] Dept of Computer Science, Guangxi University of Technology, China
[3] College of Computer and Software, Taiyuan University of Technology, China
[4] Dept of Psychology, Carnegie Mellon University, USA
[5] Dept of Life Science and Informatics, Maebashi Institute of Technology, Japan
yulinq@yahoo.com, zhong@maebashi-it.ac.jp

**Abstract.** In this paper, we present an investigation on heuristics retrieval in human problem solving by combining the computational cognitive model ACT-R (Adaptive Control of Thought-Rational) and advanced fMRI (functional Magnetic Resonance Imaging) brain imaging technique. As a new paradigm, 4*4 Sudoku is developed to facilitate this study, in which seven heuristics that can be classified into 3 groups are designed to solve two types of tasks: simple and complex ones. The cognitive processes of the two types of 4*4 Sudoku tasks are explored based on the outputs of ACT-R model. This study shows that several key elements take important roles in the retrieval of heuristics, including the ways of problem presentation, complexity of heuristics and status of goal. The fitness of model prediction to real participants' data on behavior and BOLD (Blood Oxygenation Level-Dependent) response in five predefined brain regions illustrates that our hypotheses and results are acceptable. This work is a significant step towards tackling the puzzle of the heuristics retrieval in human brain.

## 1    Introduction

Human problem solving, one of the important research issues in cognitive science and computer science especially in artificial intelligence (AI), has been studied for dozens of years [2, 7, 8, 12, 13, 15]. In their book [8], Newell and Simon observed that human being always uses heuristic strategy when solving a complex problem, such as chess. Heuristic searching strategy is also a popular method in AI, such as in solving traveling salesman problem [12], flow shops scheduling problem [7], and so on. MaCarthy pointed out that the largest qualitative gap between human performance and computer performance is in the area of heuristics [9]. One puzzle on problem solving is that how human brain retrieves and uses heuristics to speed up problem solving. The answer of this question may shed light on developing new model for Web-based information retrieving (IR), Web-based reasoning technology [16, 17]. This paper focuses on this question and

N. Zhong et al. (Eds.): BI 2009, LNAI 5819, pp. 53–62, 2009.

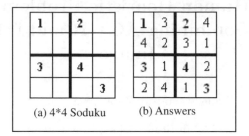

**Fig. 1.** 4*4 Sudoku and the answers

tries to examine the information processing process of human problem solving by the approach of combining computational cognitive model ACT-R (Adaptive Control of Thought-Rational) and advanced fMRI (functional Magnetic Resonance Imaging) brain imaging techniques [3, 5, 11].

In cognitive psychology and AI domains, many classical problem tasks have been used to investigate the information processing processes of problem solving, such as Tower of Hanoi, the savage and the missionary puzzle, and so on. These tasks are somewhat complicated and cost long time to solve. A good task in brain imaging experiments shall be solved in a short period with brief heuristics and be consistent for all participants. We found that the simplified 4*4 Sudoku is a good candidate with all these features. A 4*4 Sudoku, as shown in part (a) of Figure 1, is a 4*4 matrix with two mid-lines bold lines are used to divide the matrix which called boxes and some of grids are fill with digits. The task is to fill all of the empty grids so that each row, each column, and each 2*2 box contains the digits from 1 to 4 only one time each, as shown in part (b) of Figure 1. For controlling the strategies that participants use, we simplified the problem so that participants only need to find the answer in one grid which is marked by '?'.

## 2    Methods

### 2.1    Event-Related fMRI Experiment

*Tasks*. As described in Section 1, we developed a new paradigm of human problem solving, simplified 4*4 Sudoku that involves seven heuristics which can be classified into three groups.

- The first three heuristics that are most easy one involve checking in one dimension : (Row), (Column) and (Box). For example, Row heuristics is shown in part (a) of Figure 2. Three digits in the set of 1 to 4 are given in the row with grid of '?'. The answer of '?' is the resting digit.
- The second three heuristics involve checking two dimensions: (Row and Column), (Row and Box), and (Column and Box). For example, Row and Column heuristics is shown in part (b) of Figure 2. Three digits in the set of 1 to 4 are given in the row and column with grid of '?'.

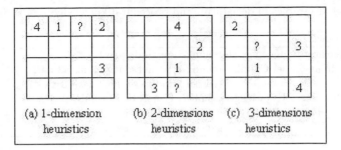

**Fig. 2.** Three types of heuristics in 4*4 Sudoku

– The last heuristics involves checking three dimensions: (Row and Column and Box), as shown in part (c) of Figure 2. Three digits in the set of 1 to 4 are given in the row, column and box with grid of '?', only one digit in each dimension.

The first three heuristics checking only one dimension are called as simple heuristics, and the resting four heuristics checking one or two more dimensions as complex heuristics.

***Materials and Procedure.*** The tasks in this experiment are to solve the problem using only one of above heuristics. Therefore, the experiment materials are two types of 4*4 Sudoku problems. The visual stimuli of the tasks also are those as shown in Figure 2. The experiment took two steps. Firstly, before fMRI scanning, all participants are trained to get familiar with the seven heuristics and to be able to use these heuristics to solve the corresponding problems. Secondly, the participants performed tasks in MRI scanner with more than 80 trials.

***Experiment Procedure.*** A trial begin with an alerting stimulus presents for two seconds followed by a 4*4 Sudoku stimulus that stays on the screen until the participant indicates they know the answer or 20 seconds has elapsed. When participants feel that they has found the answer of the problem, they press a key under the thumb of their right hands and then speak out the answer. Then the Sudoku stimulus is cleared. The ACT-R model simulates the processes from the presentation of stimulus to the key pressing.

***fMRI Scan.*** Event-related fMRI data were collected by a gradient echo planar pulse acquisition on a Siemens 3T Trio Tim Scanner. Behavioral data and fMRI brain imaging data of 16 participants were collected.

***Data Analysis and Results.*** Behavioral results RT (mean of response time over 16 participants) of the two types of 4*4 Sudoku, were calculated. Functional fMRI results, BOLD (Blood Oxygenation Level-Dependent) responses in five predefined brain regions based on [6] were also calculated.

**Table 1.** Examples of declarative and procedural knowledge in 4*4 Sudoku

| Knowledge | Example |
|---|---|
| Declarative Knowledge | (chunk-type Sudoku x1 x2 x3 x4) |
| (Chunk-Type, Chunk) | (chunk-type quesdata a1 a2 a3 a4 ... ) |
|  | (p1 isa Sudoku x1 1 x2 2 x3 3 x4 4) ... |
| Procedural Knowledge | visual-?, visual-row, visual-col, visual-box |
| (Production Rules) | encode-row, encode-col, encode-box |
|  | get-answer, pressing, ... |

## 2.2   ACT-R Model

Although some results have been obtained from fMRI data, such as BOLD response of five brain regions, these results cannot tell us detail information processing processes to understand how the problem solving is processed [14] and how these cognitive processes work on the change of BOLD response of brain regions. Therefore, ACT-R modeling is combined with fMRI data to help us solving these problems [1, 4].

***The Steps of Modeling.*** In general, the following steps are involved to set up a cognitive model in ACT-R.

- Simulating stimulus environment. This is the first step in ACT-R modeling that makes the model performing the task as a real participant.
- Defining declarative knowledge and procedural knowledge. Two knowledge definitions are the main components of an ACT-R model that help to set up a successful model.
- Setting parameters for a new model. Most of parameters are default except some of them may not fit to a new task.
- Debugging and running the model to get a prediction of behavioral data RT and functional fMRI BOLD response.

***Two Knowledge Definitions of the Model.*** In ACT-R, declarative knowledge is presented as chunk defined by chunk-type and slots while procedural knowledge is presented as production rules like if/then patterns. Table 1 lists the two knowledge definitions of 4*4 Sudoku (only part of them): chunks like the form of 1 to 4, the information related '?', production rules like encode '?' and encode heuristic information in row, column and box dimension, and so on.

***Parameters.*** Table 2 lists the parameters of the model for predicting the BOLD response in five predefined brain regions, where $m$ is the scales the magnitude of the response, $s$ is the time scale and $b$ is the exponent. Others are using default values.

***Hypotheses of Cognitive Process.*** The cognitive processes of 4*4 Sudoku problem solving are supposed as follows. Firstly, participants encode the object '?', and then they shift to related dimensions to encode the information to retrieve heuristics and omit irrelative information. While gathering the digits

**Table 2.** Parameters for predicting BOLD response of five modules

|  |  | Module Visual | Goal | Retrieval | Imaginal | Production |
|---|---|---|---|---|---|---|
|  | $m$ | 0.8 | 1 | 1 | 1.3 | 0.1 |
| Parameters | $s$ | 1.5 | 0.9 | 2.5 | 1.5 | 1.5 |
|  | $b$ | 6 | 7 | 2 | 6.7 | 6 |

related to heuristics, participants make a number of judgments by the state of goal and retrieve heuristics for the answer and finally make their responses. We simplify the processes in one sentence: only encoding valid information in related dimensions for heuristics retrieval by goal status oriented. In the model, ways of problem presentation, complexity of heuristics and status of goal take important roles in the heuristics retrieval.

## 3 Results

### 3.1 Behavioral Results

Behavioral results, the mean latencies (also called response time (RT)) over 16 participants from the presentation of tasks stimulus to the thumb press are 1.53 seconds in simple tasks and 2.90 seconds in complex tasks. These results came from behavioral data collected synchronous with fMRI scanning. The means RT are for correct trials only. The difference between simple and complex tasks is 1.37 seconds which shows that the heuristics application in complex tasks cost more time because visual encoding, memory retrieval, problem representation, goal control and production rules are more complex than simple ones. All of these cognitive elements are involved in the retrieval and application of heuristics.

### 3.2 fMRI BOLD Effect Results

Dotted lines in Figure 3 show the BOLD response of five predefined brain regions for simple and complex 4*4 Sudoku tasks. These results show that complex tasks always have larger time scale and steeper shape than the simple ones. The complexity of heuristics makes these differences. More explanation will be given later.

### 3.3 Fitting an ACT-R to the Behavioral Results

The predictions of behavioral results RT are 1.50 seconds in simple tasks and 2.84 seconds in complex tasks. The correlation of latency of ACT-R theory and real RT data is 0.999, and the difference is 0.04. The well fitness of ACT-R theory to participants data in behavioral results demonstrates that the modeling cognitive processes mentioned in Section 2, are a certain extent reasonable. Tables 3 and 4 show the information processing processes on five modules of ACT-R for the two kinds 4*4 Sudoku based on trace outputs of ACT-R model.

***Visual Module.*** For both tasks, participants need to find out '?' firstly for retrieval and then retrieve digits on related dimensions. For the simple ones, the information of heuristics is gathered only from one row (or one column/box). And visual module doesn't take much time to encode three digits because it does not need to shift focus to another dimension. The gathering digits in only one dimension make participants encode them soon, only one second, as shown in Table 3. While in complex tasks, heuristic related information disposes on at least two dimensions and costs more time to encode, and the participants have to shift focus to another dimension after encoding one. It takes about 1.5 seconds before retrieving all information for heuristics, 0.5 second more than the simple one, as shown in Table 4. Only related dimensions and valid information are encoded in the model because encoding other irrelative information will cost more and is not supported in the model. It is also supposed that the order of encoding in 4*4 Sudoku is row, column and box. Only valid information in heuristics is presented in the model.

***Goal Module.*** Before getting the answer, the state of controlling and transforming cognitive processes is changed by the complexity of heuristics. The basic procedure of control state in the goal module includes visualizing the information from input modality through eyes, encoding and integrating the information, and then retrieving the heuristics to get the answer, pressing key and speaking the answer out in the end. The complexity of heuristics differs in dimension number and digits distribution. For simple tasks, the states in the goal module are simpler than the complex ones in dimension encoding, representation, information integration and memory retrieval, as listed in Tables 3 and 4. The tasks are easy for adult participants because goal space is fixed in 1, 2, 3, 4 and only different in their orders. By goal status oriented, participants might only attend the valid information in related dimension and retrieved heuristics to answer the question.

***Retrieval Module.*** As mentioned above, the retrieval module retrieves information from declarative memory. For the simple tasks, the digits in heuristics are only in one dimension and the memory burden is least, so the retrieval module is fired one time; for the complex ones, heuristic related digits are in two or three dimensions and the brain maintained the former encoded digits. In this case, retrieval module is activated two or three times.

***Imaginal Module.*** The imaginal module in ACT-R is response for transforming problem representations, such as a change to retrieve digits from one dimension to another. So it is related to the operation of retrieval module where it is mainly determined by valid information of heuristics. Note that although there are some irrespective digits, they are not recalled, such as '3' in part (a) of Figure 2. One interpret action is that human being always attends and selects useful information actively in heuristic strategy.

***Procedural Module.*** In ACT-R architecture, the procedural module takes a role of conductor or control center in dealing with the communication of information among modules. The production rules implement the basic units of procedural knowledge. The scale of production system is decided by the

**Table 3.** Operations in five modules for simple 4*4 Sudoku (for task (a) of Fig. 2)

| Scan | Time | Visual | Goal | Retrieval | Imaginal | Production |
|---|---|---|---|---|---|---|
| 1 | 0.0 | | | | | |
| | | Encode ? | | | ? | Encode |
| | 0.5 | Encode row | Visualizing | | | Focus row |
| | | Encode 4 | Encoding | | 4 | Encode |
| | 1.0 | Encode 2, 1 | | | 2, 1 | Encode |
| | | | Integrating | | 4, 2, 1 | Solve |
| | | | Retrieving | 4, 2, 1, (3) | 3 | Retrieve fact |
| | 1.5 | | Pressing | (3) | | Press key |

**Table 4.** Operations in five modules for complex 4*4 Sudoku (for task (b) of Fig. 2)

| Scan | Time | Visual | Goal | Retrieval | Imaginal | Production |
|---|---|---|---|---|---|---|
| 1 | 0.0 | | | | | |
| | 0.5 | Encode ? | Visualizing | | ? | Encode |
| | | | | | | Focus Row |
| | | Encode row | | | | Evaluate |
| | 1.0 | Encode 3 | Encoding | | 3 | Focus Col |
| | | Encode col | | | | Encode |
| | | Encode 4 | | 3, 4, (1, 2) | 4 | |
| | 1.5 | Encode 1 | | | 1 | Evaluate |
| 2 | 2.0 | | Integrating | | 3,4,1 | Solve |
| | 2.5 | | Retrieving | 3, 4, 1, (2) | 2 | Retrieve fact |
| | 3.0 | | Pressing | (2) | | Press key |

complexity of problem. In modeling 4*4 Sudoku, complexity of heuristics, such as dimension number and valid information distribution decides the procedural system and the status of goal. With goal status controlled and oriented, procedural module organizes the schedule, including selected attention, problem representation, information communication and integration and state transformation, and so on.

## 3.4    Fitting an ACT-R to the BOLD Response

In this study, the regions of interest include five predefined regions related to five ACT-R modules [6, 10]. Solid lines in parts (a) to (e) of Figure 3 display BOLD response predictions for these regions. The horizontal coordinates in the figures are scan numbers. And the base line scans to be contrasted is two former scans before the stimulus onset. The mean correlation of ACT-R prediction and fMRI data is 0.87 and the mean difference is 0.057 which shows the prediction is acceptable. Part (a) of Figure 3 shows the results for the left fusiform region corresponding to the visual module. The predictions from the visual module show a good match in two type tasks. The correlation with the left fusiform is 0.93. The prediction of BOLD response of visual region in complex tasks shows

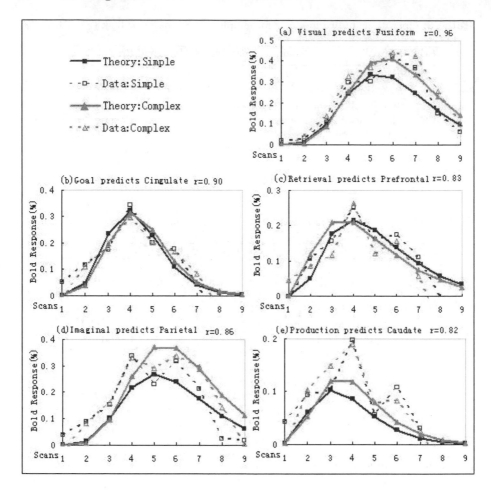

**Fig. 3.** BOLD response in ACT-R theory and fMRI data

a higher magnitude, larger time scale and steeper shapes than simple ones, just like the fMRI analysis results. Part (b) of Figure 3 shows the results for the left anterior cingulated cortex corresponding to the goal module. The correlation with the left ACC region is 0.90. Part (c) of Figure 3 shows the results for the left prefrontal region corresponding to the retrieval module. The correlation with the left prefrontal region is 0.83 which shows a less fitting. As described on declarative knowledge, the chunks only concerns four digits which is easier for an adult participant. So activity in retrieval module is lower than other modules like goal and visual module. Part (e) of Figure 3 shows the results for the left parietal region corresponding to the imaginal module. The correlation with the left parietal region is 0.86, also a receivable matching. Part (e) of Figure 3 shows the results for the left caudate region corresponding to the procedural module. The correlation with this region is 0.82, the least fitting in the model (similar to that observed by other researchers [6]). All these predictions of BOLD response show that our modeling hypotheses on cognitive processes are reasonable in some extent.

### 3.5  Summary

The simulation results in both behavioral data and BOLD response shows that the computational model of 4*4 Sudoku is reasonable and the hypotheses of modeling are explicable. The fitness of model prediction shows that some cognitive elements that affect the retrieval of heuristics. The first element is the way of problem presentation. For a given problem like 4*4 Sudoku, participants had been trained to try to get the answer with given seven heuristics. All participants knew the task environment well. In other words, the problem space for all participants is supposed the same. The way of problem presentation is simplified and the complexity of heuristics decides the cost of processes of cognition. The second one is the complexity of heuristics like the digits distribution in dimensions. This is the key factor involved in the retrieval heuristics. And the third one is the status of goal which acts a controller of the system in the whole processes, from encoding '?', retrieving answer to speaking it out. The goal status transformation is also important in heuristics retrieval.

## 4  Conclusions

Studies on human heuristic problem solving mainly involve understanding the retrieval, implication and selection of heuristics. This paper focused on the first issue (i.e. the retrieval of heuristics). Two methodologies have been used to examine the problem. The first one is ACT-R modeling based on information-processing theory and production system, and the second one is fMRI, one of brain imaging techniques. The outputs of ACT-R model prove that the cognitive hypotheses are reasonable. The good fitness of theory prediction to the real participants' data shows that the model is valid. It is obviously helpful that combining the two methods to study human brain in complex.

**Acknowledgments.** Supported by the National Natural Science Foundation of China under Grant No.60875075, Doctoral Candidates Innovative Foundation of Beijing University of Technology under Grant No.DCX-2009-065. We would like to thank Lijuan Wang efforts on fMRI experiment.

## References

1. Anderson, J.R., Bothell, D., Byrne, M.D., et al.: An Integrated Theory of Mind. Psychological Review 111(4), 1036–1060 (2004)
2. Anderson, J.R., Albert, M.V., Fincham, J.M.: Tracing Problem Solving in Real Time: fMRI Analysis of the Subject-paced Tower of Hanoi. Journal of Cognitive Neuroscience 17(8), 1261–1274 (2005)

3. Anderson, J.R., Qin, Y.L., Jung, K.J., et al.: Information-Processing Modules and Their Relative Modality Specificity. Cognitive Psychology 54(3), 185–217 (2007)
4. Anderson, J.R.: How Can the Human Mind Occur in the Physical Universe? Oxford University Press, USA (2007)
5. Anderson, J.R.: Using Brain Imaging to Guide the Development of a Cognitive Architecture. In: Gray, W.D. (ed.) Integrated Models of Cognitive Systems, pp. 49–62. Oxford University Pres, Oxford (2007)
6. Anderson, J.R., Ficham, J.M., Qin, Y.L., et al.: A Central Circuit of the Mind. Trends in Cognitive Sciences 12(4), 136–143 (2008)
7. Laha, D., Sarin, S.C.: A Heuristic to Minimize Total Flow Time in Permutation Flow Shop. Omega 37(3), 734–739 (2009)
8. Newell, A., Simon, H.A.: Human Problem Solving. Prentice-Hal, Englewood Cliffs (1972)
9. McCarthy, J.: From Here to Human-Level AI. Artificial Intelligence 171(18), 1174–1182 (2007)
10. Qin, Y.L., Sohn, M.H., Anderson, J.R., et al.: Predicting the Practice Effects on the Blood Oxygenation Level-Dependent (BOLD) Function of fMRI in a Symbolic Manipulation Task. PNAS (Proceedings of the National Academy of Sciences of the United States of America) 100(8), 4951–4956 (2003)
11. Qin, Y.L., Bothell, D., Anderson, J.R.: ACT-R meets fMRI. In: Zhong, N., Liu, J., Yao, Y., Wu, J., Lu, S., Li, K. (eds.) Web Intelligence Meets Brain Informatics. LNCS (LNAI), vol. 4845, pp. 205–222. Springer, Heidelberg (2007)
12. Renaud, J., Boctor, F.F., Ouenniche, J.: A Heuristic for the Pickup and Delivery Traveling Salesman Problem. Computers and Operations Research 27(9), 905–916 (2000)
13. Simon, H.A.: Search and Reasoning in Problem Solving. Artificial Intelligence 21, 7–29 (1983)
14. Simon, H.A.: The Information-Processing Theory of Mind. American Psychologist 50(7), 507–508 (1995)
15. Stocco, A., Anderson, J.R.: Endogenous Control and Task Representation: An fMRI Study in Algebraic Problem Solving. Journal of Cognitive Neuroscience 20(7), 1300–1314 (2008)
16. Zhong, N., Liu, J., Yao, Y., Wu, J., Lu, S., Qin, Y., Li, K., Wah, B.W.: Web Intelligence Meets Brain Informatics. In: Zhong, N., Liu, J., Yao, Y., Wu, J., Lu, S., Li, K. (eds.) Web Intelligence Meets Brain Informatics. LNCS (LNAI), vol. 4845, pp. 1–31. Springer, Heidelberg (2007)
17. Zhong, N., Liu, J.M., Yao, Y.Y.: Envisioning Intelligent Information Technologies through the Prism of Web Intelligence. Communications of the ACM 50(3), 89–94 (2007)

# EEG/ERP Meets ACT-R: A Case Study for Investigating Human Computation Mechanism

Shinichi Motomura, Yuya Ojima, and Ning Zhong

Maebashi Institute of Technology
460-1 Kamisadori-Cho, Maebashi-City 371-0816, Japan
motomura@maebashi-it.org, zhong@maebashi-it.ac.jp

**Abstract.** EEG (electroencephalograph) provides information about the electrical fluctuations between neurons that characterize brain activity, and measurements of brain activity at resolutions approaching real time. On the other hand, cognitive architectures such as ACT-R would explain how all the components of the mind work together to generate coherent human cognition. Thus EEG/ERP (event-related potential) and ACT-R will provide two aspects to explore the cognitive processes and their neural basis. In this paper, we present a case study by combining EEG/ERP and ACT-R for investigating human computation mechanism. In particular, we focus on two digits addition tasks with or without carry, and systematically perform a set of behavior and EEG experiments, as well as with the help of ACT-R simulation. Preliminary results show the usefulness of our approach.

## 1 Introduction

Multi-aspect analysis and simulation is an important methodology in Brain Informatics, which emphasizes on a *systematic* way for investigating human information processing mechanisms [18,20,21]. The existing results, as reported over the last few decades about understanding human information processing mechanism, are greatly related to progress of measurement and analysis technologies. EEG (electroencephalograph) is a popular, relatively handy measurement technique that provides information about the electrical fluctuations between neurons that characterize brain activity, and measurements of brain activity at resolutions approaching real time. Technologies have been developed to check the synchronous neural oscillations from EEG/ERP (event-related potential) data, which may reveal the dynamic cognitive processes in human brain [16]. On the other hand, cognitive architectures such as ACT-R (Adaptive Control of Thought - Rational) would explain how all the components of the mind work together to generate coherent human cognition [1,2], which is also a kind of effort towards conceptual modeling of the brain. Thus EEG/ERP and ACT-R will provide two aspects to infer the cognitive processes and their neural basis. Hence, combining EEG/ERP and ACT-R might be an appropriate approach to help us to reach the goal of investigating the spatiotemporal feature and flow of human information processing system (HIPS) and investigating neural structures and neurobiological processes related to the activated areas.

N. Zhong et al. (Eds.): BI 2009, LNAI 5819, pp. 63–73, 2009.

In this paper, we present a case study by combining EEG/ERP and ACT-R for investigating human computation mechanism. In particular, we focus on two digits addition tasks with two difficulty levels: with or without carry, and systematically perform a set of behavior and EEG experiments, as well as with the help of ACT-R modeling, to explore the cognitive processes and their neural basis. The rest of the paper is organized as follows. Section 2 provides background and related work on cognitive neuroscience, human thinking process and ACT-R research. Sections 3 explains how to design the experiments of an ERP mental arithmetic task. Sections 4 describes how to do ERP/topography data analyses and ACT-R simulation, respectively, as an example to investigate human computation mechanism and to show the usefulness of combining ERP and ACT-R. Finally, Sections 5 gives concluding remarks.

## 2  Background and Related Work

Generally speaking, capabilities of human intelligence can be broadly divided into two main aspects: perception and thinking. So far, the main disciplines with respect to human intelligence are cognitive psychology that mainly focuses on studying mind and behavior with respect to "thinking oriented" high cognitive functions and models, as well as neuroscience that mainly focuses on studying brain and biological models of intelligence. In cognitive neuroscience, although many advanced results with respect to "perception oriented" study have been obtained, only a few of preliminary, separated studies with respect to "thinking oriented" and/or a more whole information process have been reported [3,6,14]. Furthermore, an integrated interpretation with those results from both cognitive psychology and cognitive neuroscience is needed.

ACT-R, one of the best known, fully implemented, and free to public cognitive architecture models, is a theory on how the structure of brain achieves the function of adaptive cognition in a system level, as well as a platform to build computational models to simulate/predict human cognitive behavior, including a wide cognitive problems. It can perform complex dynamic tasks, which usually emphasize perceptual-motor components and their coordination with other cognitive components (learning and memory, reasoning, and so on) and with strong time pressure [1,2,13].

In recent years, a series of fMRI experiments have been performed to explore the neural basis of cognitive architecture and to build a two-way bridge between the information processing model and fMRI [11,12]. The patterns of the activations of brain areas corresponding to the buffers of the major modules in ACT-R are highly consistent across these experiments; and ACT-R has successfully predicted the Blood Oxygenation Level-Depend (BOLD) effect in these regions. A recent result is to use the imaging data to identify mental states as the student is engaged in algebraic problems solving, in which fMRI is used to track the learning of mathematics with a computer-based algebra tutor and cognitive models in the ACT-R architecture is used to interpret imaging data [1].

Problem-solving is one of main capabilities of human intelligence and has been studied in both cognitive science and AI [10], where it is addressed in conjunction with reasoning centric cognitive functions such as attention, control, memory, language, reasoning, learning, and so on. We need to better understand how human being does complex adaptive, distributed problem solving and reasoning, as well as how intelligence evolves for individuals and societies, over time and place [5,8,15,17,19,21].

As a step in this direction, a series of EEG/ERP experiments have been performed to explore the neural basis of human computation processing [9]. It includes the mental arithmetic tasks with visual stimuli for obtaining EEG/ERP human brain data as an example to investigate human problem solving process. ERP is a light and sound brain potential produced with respect to the specific phenomenon of spontaneous movement [4]. In addition to using the traditional analysis tools, we have observed that EEG/ERP data are *peculiar* with respect to a specific state or the related part of a stimulus by using an agent-enriched peculiarity oriented mining (A-POM) as a knowledge discovery approach that automates EEG/ERP data analysis and understanding [21]. Also to our best knowledge, this study is the first time to combine EEG/ERP with ACT-R.

# 3 Experimental Design of Mental Arithmetic Tasks

## 3.1 Outline of Experiments

Figure 1 gives an example of computation processing from the macro viewpoint, which consists of several component functions of the human computation mechanism, including attention, interpretation, short-term memory, understanding of work, computation, and checking. Based on the brain informatics methodology, the data obtained from a cognitive experiment and/or a set of cognitive experiments expect to be used for multi-purpose. For example, it is possible that our experiments meet the following requirements: investigating the mechanisms of human visual and auditory systems, computation, problem-solving (i.e. the computation processing is regarded as an example of problem-solving processes), and spatiotemporal characteristics and flow of HIPS in general. Furthermore, we set multiple difficulty levels with respect to a series of computation tasks in a set of experiments.

More specifically, the task of our cognitive experiments is to show a numerical calculation problem to a human subject, and to ask the subject to solve it in mental arithmetic. The numerical calculation used in the experiments is the addition problem with the following form: "augend + addend = sum" for *on-task* that is the state on which the human subject is calculating by looking at a number. Furthermore, we also need to define another state called *off-task* on which the human subject is looking at the number that appears at random.

Figure 2 shows an example of the screen state transition. For the current case study, we set two difficulty levels: level 1 is two digits addition without carry and level 2 is two digits addition with carry, in order to investigate the

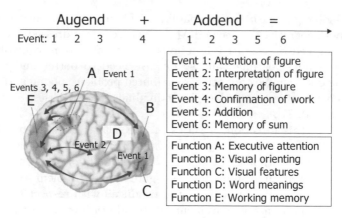

**Fig. 1.** Human computation processing

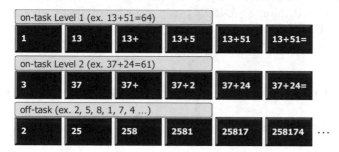

**Fig. 2.** An experimental design with two difficulty levels

spatiotemporal features and flow of the computation processing, as well as its neural basis by comparing in different difficulty levels.

## 3.2 Trigger Signal, Timing Diagram and Measurement Conditions

In order to derive ERP data, the trigger signal is fired with an interesting event since it is necessary to obtain ERP by the addition average based on the signal after measured. In this study, since we pay attention to each event of "augend", "plus", "addend", and "equal" presentations in computation activities, the pre-trigger was set to 200 msec, and an addition between two digits is recorded in 2000 msec, respectively. Figure 3 provides an example of the timing diagram, in which "Au" denotes an augend, "Ad" denotes an addend, and "Rn" denotes a random number. Furthermore, for the addition with 2 digits, "Au 1" and "Ad 1" are the most significant digit (MSD) of augend, and "Au 2" and "Ad 2" are the least significant digit (LSD) of augend, respectively. Moreover, the number from Rn 1 to Rn 8 in the off-task is 1-digit random number.

The EEG activity was recorded using a 128 channel BrainAmp amplifier (Brain Products, Munich, Germany) with a 128 electrode cap. The electrode cap is based on an extended international 10-20 system. Furthermore, eye

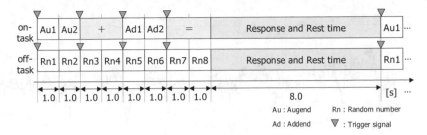

**Fig. 3.** Timing diagram of all tasks

movement measurement (2ch) is also used. The sampling frequency is 1000 Hz. The number of experimental subjects is 7.

## 4   Data Analysis and Simulation

The same mental arithmetic tasks as described in Section 3 are used to combine ERP/potential topography analyses and ACT-R simulation from the viewpoint of spatiotemporal characteristics. This is because ERP analysis provides an excellent temporal resolution. The advantage of potential topography analysis is that it catches the electrical fluctuation and gives a rough visualization of the spatiotemporal characteristics for recognizing distribution and appearance of positive and negative potentials. Furthermore, ACT-R simulation can predict the process of information processing in a 5 msec scale and can specify activated areas by modularized units for the real brain ERP data. By using such a multi-aspect approach, we attempt understand human problem solving process in depth.

### 4.1   ERP Data Analysis

For the measured EEG data, a maximum of 102 addition average processing were performed, and the ERPs were derived by using Brain Vision Analyzer (Brain Products, Munich, Germany). Generally speaking, the Wernicke area of a left temporal lobe and the prefrontal area are related to the computation process [7]. In this study, we pay attention to recognition of the number, short-term memory and integrated processing, vision, as well as comparing Level 1 and Level 2 by focusing on the following three important channels: AFz, CP3, and Oz.

Figure 4 shows the ERP with respect to the three important channels: AFz, CP3, and Oz. First of all, from the on-task related waves, some remarkably results are found in each stage of the computation process, such as augend, plus, addend, and equal. For instance, it can be confirmed that P250 appears in channel AFz, P300 or P400 appears in CP3 that is near the part of language interpretation, and N250 appears in the visual projection area Oz. Furthermore, although it is large in the stages of augend and addend presented, the reaction is small in the stage of the equal sign presented.

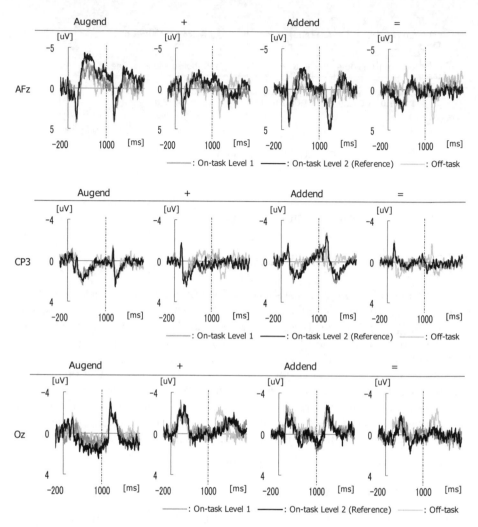

**Fig. 4.** ERP comparison among on-task Level 1, on-task Level 2 and off-task

It is suggested that the computing activity is started at once when an addend is presented. Moreover, there is a little difference between Level 1 (without carry) and Level 2 (with carry) from a global view. However, the delay is caused in Level 2 with carry about 25 msec in the stages of addend and equal presented. On the other hand, in comparison between on-task and off-task, both amplitude and timing are resembled in the first half of the computation process, but it is also confirmed that the latent time of off-task is short when entering the latter half of the computation process. It is guessed that this is related to some attention to the number related cognition. We can also see that comparing with off-task, there is the latency of reaction with 100 to 200 msec in the peak when addend and equal are presented in on-task.

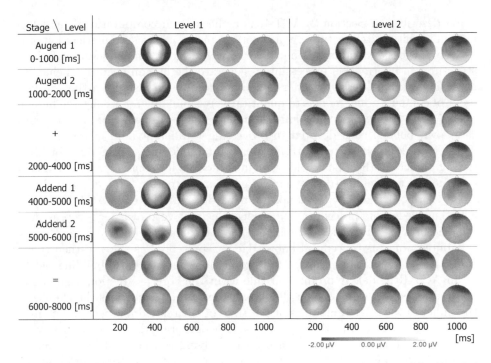

**Fig. 5.** The spatiotemporal feature of a computation process represented by Level 1 and Level 2 topographies

## 4.2 Potential Topograph Analysis

Figure 5 shows the spatiotemporal feature of a computation process represented by Level 1 and Level 2 topographies. Different from ERP, the potential topographies use only the presentation trigger signal in augend presented. We obtain these topographies by adding the average of seven subjects using an ordinary EEG analysis tool. In this research, we focus on investigating the difference between Level 1 and Level 2 in the computation process with respect to 200 msec intervals at the computation stages for a task. Although both topographies denote noticeable potentials with respect to augend or addend presented, it can be confirmed that the potential distribution is with some difference between these topographies, in particular, in the equal sign stage. According to testee questionnaires, the computing time is enough because Level 1 is easy. Therefore, it is guessed that the subjects were able to do the confirmation computation in the last stage.

## 4.3 ACT-R Simulation

ACT-R is applied to the two digits computation problem for deeply understanding the meaning of the results obtained by ERP and topographies. In particular, comparing the computation processes in Level 1 (without carry) and Level 2 (with carry). In this study, a present latest version ACT-R 6 was used.

As described in Section 2, ACT-R is a theory and computational model of human cognitive architecture. As a theory, it proposes the systematical hypothesis on the basic structure of human cognitive system and functions of these structures in information processing to generate the human cognitive behavior; as a computational model, it offers a computer software system for the development of computational models to quantitatively simulate and predict the human behavior for a wide range of cognitive tasks. It consists of a set of modules with their own buffers, each devoted to processing a different kind of information. These modules include a declarative memory module, a visual module for identifying objects in the visual field, a manual module for controlling the hands, a goal module for keeping track of current goals and intentions, and other modules.

Furthermore, there are two kinds of knowledge represented in ACT-R, declarative knowledge and procedural knowledge. Declarative knowledge is represented in structures called chunks with chunk-types as categories and slots as category attributes, whereas procedural knowledge is represented as rules (if-then rules) called productions. Thus chunks and productions are the basic building blocks of an ACT-R model. A production rule as in the core production system can be fired based on the chunks in these buffers and then it changes the chunks in the buffer of the related modules or the state of the related modules, which may leads to fire a new production rule and so on to generate the cognitive behavior.

In cognitive psychology, human problem solving can be modeled as operating on the human heart symbol (the form of information representation in the mind), and retrieving memory and managing states of the goal. Based on the observation, the following chunks were defined in the present study.

- Chunk that shows addition knowledge of one digit
- Chunk for carrying cognition
- Chunk for controlling states of a goal
- Chunk that maintains mental representation.

Furthermore, production rules were defined by dividing the visual perception process and the computing process. In each step of operations, states of the goal are managed, the mental representation is operated, the memory is retrieved, and it demands from the buffer of an intellectual module. The visual perception process was divided into the following steps.

- Position tracking for the displayed number
- Movement of attention to the detected position
- The number is recognized and maintained as a mental representation.

The computation process was divided into the following steps.

- Addition of units digit
- Determination of carrying
- Addition of tens digit
- Processing at the case of carrying
- Collation and input of results.

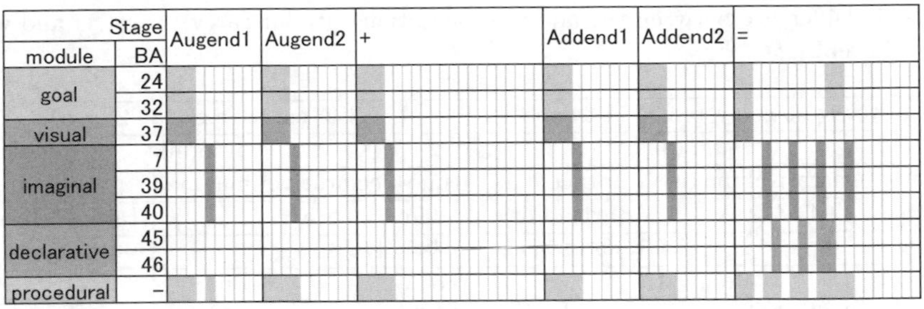

| module | Stage BA | Augend1 | Augend2 | + | | Addend1 | Addend2 | = |
|---|---|---|---|---|---|---|---|---|
| goal | 24 | | | | | | | |
| | 32 | | | | | | | |
| visual | 37 | | | | | | | |
| imaginal | 7 | | | | | | | |
| | 39 | | | | | | | |
| | 40 | | | | | | | |
| declarative | 45 | | | | | | | |
| | 46 | | | | | | | |
| procedural | – | | | | | | | |

**Fig. 6.** Results of ACT-R Simulation for on-task Level 2 (with carry)

Figure 6 shows the ACT-R simulation result of two digits addition problem for on-task Level 2 (with carry). The vertical axis denotes the modules related to the computation activities in ACT-R, and the horizontal axis denotes the time-series of the computation activities.

The modules are fired when computing are declarative, imaginal, and procedural ones. It is suggested that the region BA (Brodmann Area) 45, BA 46, BA 7, BA 39, BA 40 and basal ganglia corresponding to these modules deeply contribute to such a computation related problem solving process. When augend, operator, and addend are presented, the displayed position tracking to the number and the attention shift are reflected and the visual module is activated. Moreover, to maintain the recognized number as a mental representation, the imaginal module is fired. It is suggested that the addition knowledge is referred to the declarative module when the equal sign is presented, and the declarative memory (semantic memory) contributes to the problem solving. When a sum is presented, both the manual module and the motor area are remarkable. The goal module acts through the whole, and the process that arrives at the state of the goal while managing a small step is suggested. Moreover, the procedural module that plays the role to control the activity of all modules also acts through the whole.

### 4.4 A Comparative Study of Results

We focus on deeply investigating the delay problem that is caused by the difference between the addition operation without carry (Level 1) and with carry (Level 2). In the ERP analysis, it can be confirmed that, compared with Level 1, there exists a delay about 25 msec in Level 2. However, it is difficult for a more detailed analysis since the inconsistency of time and individual difference of subjects. For this reason, we individually investigated the subjects who can be classified into 2 types: good and not good at computation. The results shows that the delay is hardly seen in the subjects who are good at computation, but the delay about 50 msec to 100 msec appears in the subjects who are not good at computation.

The output of an ACT-R simulation is a time course for when and how long the activations of each module involved in the task. Based on it, one can predict the performance of the subjects (such as reaction time and accuracy) as a general model. Therefore, we can use such a general model as a basis and combine with ERP/topography to analyse individual difference of subjects.

**Table 1.** Difference between the addition operation without carry (Level 1) and with carry (Level 2)

| Level 1: no-carry (e.g. 24+34=58) | | | | Level 2: carry (e.g. 13+28=41) | | | |
|---|---|---|---|---|---|---|---|
| | module | | | | module | | |
| time | declarative | imaginal | procedural | time | declarative | imaginal | procedural |
| 6.435 | 4+4=? | | start-addition | 6.435 | 3+8=? | | start-addition |
| 6.485 | | | add-ones | 6.485 | 11=10+? | | add-ones |
| 6.685 | | one=8 | | 6.685 | | one=1 | |
| 6.735 | 2+3=? | | add-tens | 6.735 | 1+2=? | | add-tens |
| 6.935 | | ten=5 | | 6.935 | | ten=3 | |
| 6.985 | | | | 6.985 | 3+carry=? | | add-carry |
| 7.035 | | ans=58 | | 7.035 | | | |
| 7.085 | | | | 7.085 | | | ans=41 |

Table 1 shows the results of ACT-R simulation with respect to Level 1 (without carry) and Level 2 (with carry). Comparing the processes, we can see that the declarative memory is accessed in the case with carry, and the delay about 50 msec appears (step-11 in the ACT-R result). This is because a carry is caused by the addition operation of "3" and "8" of the units digit, and the carry will be operated with "3" of the sum of the tens digit, which is more complex than the computation in Level 1.

## 5    Conclusion

In this paper, we presented an approach of combining EEG/ERP analysis and ACT-R simulation to understand deeply a two digit mental computation problem as a case study for investigating human computation mechanism. It was confirmed that the retardation phenomenon of the reaction appears in the addition operation with carry, compared with the addition operation without carry. And individual difference in performance was also discussed. The preliminary results show the usefulness of our approach.

Our future work includes extending our approach to combine with fMRI since EEG topographies have limitation in spatial resolution and need to combine with fMRI data analysis for identifying the activated areas more accurately. Thus, combining EEG/ERP, fMRI and ACT-R will provides an powerful approach to help us to investigate the spatiotemporal feature and flow of HIPS, and its neural structures and neurobiological processes related to the activated areas.

## References

1. Anderson, J.R., Bothell, D., Byrne, M.D., Douglass, S., Lebiere, C., Qin, Y.: An Intergrated Theory of the Mind. Psychological Review 111(4), 1036–1060 (2004)
2. Anderson, J.R.: How can the Human Mind Occur in the Physical Universe. Oxford University Press, Oxford (2007)

3. Gazzaniga, M.S. (ed.): The Cognitive Neurosciences III. The MIT Press, Cambridge (2004)
4. Handy, T.C.: Event-Related Potentials, A Methods Handbook. The MIT Press, Cambridge (2004)
5. Liu, J., Jin, X., Tsui, K.C.: Autonomy Oriented Computing: From Problem Solving to Complex Systems Modeling. Springer, Heidelberg (2005)
6. Megalooikonomou, V., Herskovits, E.H.: Mining Structure-Function Associations in a Brain Image Database. In: Cios, K.J. (ed.) Medical Data Mining and Knowledge Discovery, pp. 153–179. Physica-Verlag (2001)
7. Mizuhara, H., Wang, L., Kobayashi, K., Yamaguchi, Y.: Long-range EEG Phase-synchronization During an Arithmetic Task Indexes a Coherent Cortical Network Simultaneously Measured by fMRI. NeuroImage 27(3), 553–563 (2005)
8. Mitchell, T.M., Hutchinson, R., Niculescu, R.S., Pereira, F., Wang, X., Just, M., Newman, S.: Learning to Decode Cognitive States from Brain Images. Machine Learning 57(1-2), 145–175 (2004)
9. Motomura, S., Hara, A., Zhong, N., Lu, S.: An Investigation of Human Problem Solving System: Computation as an Example. In: Kryszkiewicz, M., Peters, J.F., Rybiński, H., Skowron, A. (eds.) RSEISP 2007. LNCS (LNAI), vol. 4585, pp. 824–834. Springer, Heidelberg (2007)
10. Newell, A., Simon, H.A.: Human Problem Solving. Prentice-Hall, Englewood Cliffs (1972)
11. Qin, Y., et al.: Predicting the Practice Effects on the Blood Oxygenation Level-dependent (BOLD) Function of fMRI in a Symbolic Manipulation Task. PNAS 100, 4951–4956 (2003)
12. Qin, Y., Bothell, D., Anderson, J.R.: ACT-R meets fMRI. In: Zhong, N., Liu, J., Yao, Y., Wu, J., Lu, S., Li, K. (eds.) Web Intelligence Meets Brain Informatics. LNCS (LNAI), vol. 4845, pp. 205–222. Springer, Heidelberg (2007)
13. Sohn, M.-H., Douglass, S.A., Chen, M.-C., Anderson, J.R.: Characteristics of Fluent Skills in a Complex, Dynamic Problem-solving Task. Human Factors 47(4), 742–752 (2005)
14. Sommer, F.T., Wichert, A. (eds.): Exploratory Analysis and Data Modeling in Functional Neuroimaging. The MIT Press, Cambridge (2003)
15. Sternberg, R.J., Lautrey, J., Lubart, T.I.: Models of Intelligence. American Psychological Association (2003)
16. Ward, L.M.: Synchronous Neural Oscillations and Cognitive Processes. TRENDS in Cognitive Sciences 7(12), 553–559 (2003)
17. Zadeh, L.A.: Precisiated Natural Language (PNL). AI Magazine 25(3), 74–91 (Fall 2004)
18. Zhong, N., Wu, J.L., Nakamaru, A., Ohshima, M., Mizuhara, H.: Peculiarity Oriented fMRI Brain Data Analysis for Studying Human Multi-Perception Mechanism. Cognitive Systems Research 5(3), 241–256 (2004)
19. Zhong, N.: Impending Brain Informatics Research from Web Intelligence Perspective. International Journal of Information Technology and Decision Making 5(4), 713–727 (2006)
20. Zhong, N., Liu, J., Yao, Y., Wu, J., Lu, S., Li, K. (eds.): Web Intelligence Meets Brain Informatics. LNCS (LNAI), vol. 4845, pp. 1–31. Springer, Heidelberg (2007)
21. Zhong, N., Motomura, S.: Agent-Enriched Data Mining: A Case Study in Brain Informatics. IEEE Intelliegnt Systems 24(3), 38–45 (2009)

# Evaluation of Probabilities and Brain Activity - An EEG-Study

Ralf Morgenstern[2,*], Marcus Heldmann[1], Thomas Münte[1], and Bodo Vogt[2]

[1] Otto-von-Guericke-University Magdeburg, Neuropsychology, Universitätsplatz 2,
P.O. Box 4120, 39106 Magdeburg, Germany
[2] Otto-von-Guericke-University Magdeburg, Faculty of Economics and Management,
Universitätsplatz 2, P.O.Box 4120, 39106 Magdeburg, Germany
ralf.morgenstern@ww.uni-magdeburg.de

**Abstract.** This paper focuses on the problem of probability weighting in the evaluation of lotteries. According to Prospect Theory a probability of 0.5 has a weight of smaller than 0.5. We conduct an EEG experiment in which we compare the results of the evaluation of binary lotteries by certainty equivalents with the results of the bisection method. The bisection method gives the amount of money that corresponds to the midpoint of the utilities of the two payoffs in a binary lottery as it has been shown previously. In this method probabilities are not evaluated. We analyzed EEG data focused on whether a probability is evaluated or not. Our data show differences between the two methods connected with the attention towards sure monetary payoffs, but they do not show brain activity connected with a devaluation of the probability of 0.5.

## 1  Introduction and Theory

In risky decision making Prospect Theory [1, 2] is one of the most accepted theories. One central point in Prospect Theory is probability weighting. One key aspect of the probability weighting function is that the probability of 0.5 has a weight smaller than 0.5. We design an EEG study in which we try to find brain activity associated with the perception of the probability of 0.5. In a previous EEG study [3] the bisection method as a direct method without risk was compared with the certainty equivalent method as an indirect method that is connected with risk. Both methods are used for eliciting a utility function and this study revealed that the bisection method is suitable as a reference method for eliciting utility functions. We use these two methods to isolate the effect of the evaluation of the probability of 0.5 in the experimental design and in the EEG data.

In the following part we shortly describe both methods and the problem in more detail. Then we present the experimental part with our results and the conclusion.

### 1.1  Certainty Equivalent Method

The certainty equivalent method elicits a utility function by determining certainty equivalents of lotteries in which the payoffs $x^-$ and $x^+$ occur with a probability of

---

* Corresponding author.

N. Zhong et al. (Eds.): BI 2009, LNAI 5819, pp. 74–83, 2009.
© Springer-Verlag Berlin Heidelberg 2009

$p = 0.5$. Based on decision of a subject between a binary lottery and a sure payoff, the certainty equivalent CE represents the sure payoff when the subject becomes indifferent between the lottery and the sure payoff.

## 1.2 Bisection Method

Applying this method, a utility function is elicited by determining the amount of money CU that corresponds to the midpoint of the utilities of the two amounts of money $x^-$ and $x^+$. During the experiment the subjects are asked to evaluate their perceived 'happiness that money brings' [4] in order to achieve a monetary valuation without using lotteries. This study uses the term 'joy' at receiving an amount of money when applying the bisection method in order to provide the subjects with a monetary valuation context (see figure 1).

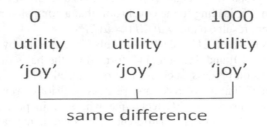

**Fig. 1.** Schematic presentation of the bisection method for $x^- = 0$ and $x^+ = 1000$

## 1.3 Theory: Does CU Equal CE?

After comparing the prediction of models of decision theory for both methods it can be stated that CU will equal CE does not apply for all theories. While $CU = CE$ can be reasoned by Expected Utility Theory, the same is not true for Prospect Theory [1, 2] due to its use of a Probability Weighting Function.

If the utility function in both methods is normalized to 0 and 1 and the offered lottery of the CE method has a fifty-fifty chance ($p = 0.5$), a theoretical comparison can be made as follows:

The bisection method asks for the same difference of 'joy' between three outcomes $x^-$, $CU$ and $x^+$. This is achieved by perceiving $CU$ as the monetary amount corresponding to the midpoint of joy between $x^-$ and $x^+$.

As far as normalization is concerned, the utility of $x^-$ and $x^+$ is $u(x^-) = 0$    and $u(x^+) = 1$. Thus, the utility of the perceived midpoint $CU$ is

$$u(CU) = \frac{u(x^-)+u(x^+)}{2} = \frac{1}{2}.$$

The offered lottery of the CE method has two values $x^-$ and $x^+$ as in the bisection method. According to Expected Utility Theory, the utility of the certainty equivalent is equal to the probability p of the fifty-fifty lottery, since

$$u(CE)_{EUT} = u(x^+) \times p + u(x^-) \times (1 - p) = 0.5.$$

Hence, the perceived midpoint of the difference of joy has the same utility as the certainty equivalent of an indifference decision between a fifty-fifty lottery and a sure payment. It follows that a comparison of these two methods will not yield different results if Expected Utility Theory holds.

The CE method as used for the development of Prospect Theory differs in this comparison, given that the utility of an indifference decision between a sure outcome and a lottery does not equal the given probability p as a result of the weighting function, consequently

$$u(CE)_{PT} = u(x^+) \times w(p) + u(x^-) \times (1 - w(p)) = w(0.5).$$

The Probability Weighting Function is stated to be inverse S-shaped, with a probability of 50 % being affiliated to a section of underweighting probabilities [2, 5, 6]. Further studies [7, 8] also confirmed that the crossover point from an overweighting to an underweighting of probabilities is about $p = 0.35$. As a result, the value of $w(0.5)$ is actually lower than 0.5, leading to the conclusion that a comparison of these two methods yields different results, namely $u(CU) > u(CE)$.

Different results between CU and CE would also be expected by theories of Regret [9], Disappointment [10] and Tension [11], formed on the basis of additional effects throughout the decision process, such as emotional reactions etc. In order to investigate the neural underpinnings of risky and riskless decisions event-related brain potentials (ERPs) were used, an EEG technique where brain potentials are recorded time-locked to external stimuli [12]. Different ERP components can be characterized by three criteria: time, place and polarity of the components' appearance. The P300 (or P3), a positive deflection at centro-parietal electrode sites peaking at 300 ms or more after the stimulus' presentation, is basically related to the cognitive processes like working memory, allocation of attentional resources, stimulus novelty and the stimulus' dependence on a given task [13-16]. Other cognitive processes having an impact on the P300 variability like stimulus frequency [13, 14], emotional value [14, 17, 18] or the stimulus' relevance can be traced back to the aforementioned concepts.

Coming back to the economical question that CU equals CE, the P300 can be utilized to reveal processes that are not reflected in observable behavior. For example, according to Prospect Theory the information of performing a lottery should result in a devaluation of the amount used in this task. But when does this devaluation take place? And is this devaluation reflected in the processing of the information when providing stimuli? In the present paper the focus is directed on the attention-capturing processes of the bisection method and the certainty equivalent method. The expected result for the riskless method would be less attention allocation processes compared to the method that is connected with risk.

## 2 Material and Methods

### 2.1 Participants

16 right-handed (9 women) and neurologically healthy subjects participated in this study after giving informed consent. They were paid 7 Euro per hour for their participation.

## 2.2 Experimental Procedure

The subjects took part in two sessions which were scheduled two weeks apart. They were seated in a comfortable chair in front of a 19 inch CRT display. Each session started with a test block to familiarize the subjects with the task. It was the subject's task to make decisions by clicking two mouse buttons with their left or right index finger.

First, two numbers ($x^-$ and $x^+$) were shown to the subject within a white frame (see figure 2), which turned into one out of two colors, either light blue or pink. According to this color codification, the subjects distinguished between two conditions: the CE method termed as 'lottery condition' and the bisection method termed as 'bisection condition'. Then a third number was revealed, located between the numbers shown before, and the subjects had to make a decision by means of a YES/NO response. Finally the screen turned back into black.

As for the lottery condition, the outer numbers represented payoffs of a fifty-fifty lottery and the inner number a sure payoff. The subjects were asked whether they preferred a sure payoff (YES), or opted for playing the fifty-fifty lottery (NO).

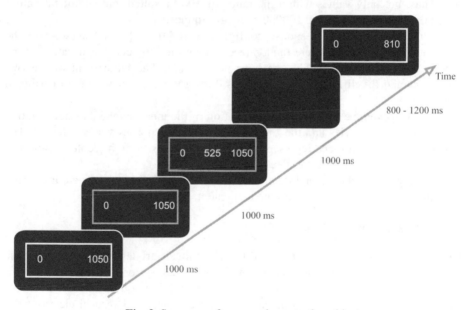

**Fig. 2.** Sequence of screens shown to the subject

Additionally, for the bisection condition the outer numbers corresponded to the utility interval boundaries, while the inner number characterized the perceived utility interval center. The subjects had to decide whether they sensed the first utility interval as larger than the second utility interval (YES or NO) concerning the perceived joy at receiving these amounts of money.

Hence, the question in both conditions resulted in asking whether the same interval had been perceived as larger or not (see figure 3).

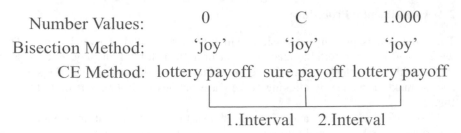

**Fig. 3.** Comparison of the experimental question for the CE method and the bisection method

Furthermore, identical numerical values were presented in both conditions. The left number $x^-$ was constantly zero, while the right number $x^+$ included 15 different values that were placed around 1000. The inner number varied between seven categories, containing the exact arithmetic mean C of the two outer numbers as well as ranges of 50 units (C+50, C-50), 150 units (C+150, C-150) and 300 units (C+300, C-300) from the center. All numerical values were multiplied by the factors 1, 10 and 100. Thus, not only values within the range of 1000 resulted, but additional values within the range of 10000 and 100000 were also induced.

Color codification and response configuration of the YES/NO answers to the right and left index finger were randomized among the subjects. The duration of the experiment was approximately one hour per session. The adjustment of the two conditions and the different number values during the experiment were distributed randomly.

In order to examine the stimulus presentation of the inner number, a fragmentation into the seven categories and the following YES/NO responses was conducted. The NO answers, in this respect, represent negative deflections from the center position C, and the YES answers positive deflections.

For analyzing the behavioral data only the averaged YES answers were taken into account, given that the NO answers are equivalent.

## 2.3 EEG Recording and Analysis

The electroencephalogram was recorded from 29 thin electrodes mounted in an elastic cap and placed according to the international 10-20 system. The EEG was re-referenced offline to the mean activity at the left and right mastoid. In order to enable the offline rejection of eye movement artifacts, horizontal and vertical electrooculo-grams (EOG) were recorded using bipolar montages. All channels were amplified (bandpass 0.05 – 30 Hz) and digitized with 4 ms resolution, impedances were kept below 10 kΩ.

EEG-data with eye blinks were corrected using a blind source separation method [19]. EEG-periods with pulse artifacts were excluded from the data set. Following the artifact procedure, stimulus-locked bins were calculated (epoch length 900 ms, baseline 100 ms) for each subject and the following conditions (see table 1).

**Table 1.** Stimulus-locked bins for the EEG data

| Condition | Inner Number Category | Followed Response | Bin |
|---|---|---|---|
| Lottery Condition | C-300 | No | C-300NoLott |
| Lottery Condition | C-150 | No | C-150NoLott |
| Lottery Condition | C-50 | No | C-50NoLott |
| Lottery Condition | Center | No | CenterNoLott |
| Lottery Condition | Center | Yes | CenterYesLott |
| Lottery Condition | C+50 | Yes | C+50YesLott |
| Lottery Condition | C+150 | Yes | C+150YesLott |
| Lottery Condition | C+300 | Yes | C+300YesLott |
| Bisection Condition | C-300 | No | C-300NoBisec |
| Bisection Condition | C-150 | No | C-150NoBisec |
| Bisection Condition | C-50 | No | C-50NoBisec |
| Bisection Condition | Center | No | CenterNoBisec |
| Bisection Condition | Center | Yes | CenterYesBisec |
| Bisection Condition | C+50 | Yes | C+50YesBisec |
| Bisection Condition | C+150 | Yes | C+150YesBisec |
| Bisection Condition | C+300 | Yes | C+300YesBisec |

## 3  Results

### 3.1  Behavioral Data

A within-subject analysis of variance (ANOVA) was performed for the behavioral data, with the condition, the inner number category and the scaled factor serving as inner subject factors. The ANOVA revealed no significant difference between both conditions ($F=0.307$, $df=1$, $p=0.588$). Further verification of both conditions using a paired t-test resulted in no significant values on a 5 % significant level (see table and figure 8 in the Appendix). Furthermore, a significant value is existent for the scaled factors ($F=1.313$, $df=6.762$, $p=0.012$), but not for other factor interactions (see table 3 in the Appendix).

### 3.2  EEG Data

The figures 4 and 5 show stimulus-locked ERPs at the CZ electrode for the bins described in the previous section. Most pronounced differences are located about 500 ms poststimulus. ERPs related to the center position (C-50, C, C+50) hardly deviate, whereas the categories C-300, C-150, C+150 and C+300 tend to have a larger positivity. Obviously, the most pronounced positivity is related to the bin in the lottery condition.

ERP waveforms were analyzed by a set of ANOVAs, with the focus being directed on the mean amplitudes in different time periods and differentiated by the lateral, parasagital and midline location as well as by the followed response.

A high significant difference could be determined at a time period of 450-600 ms poststimulus (p<0.02, see table 5 and 6 in the Appendix) for the inner number categories. Based on a 5% significant level, no significant differences exist when regarding both conditions as within-subject factor (see table 5 and 6 in the Appendix); neither for the YES answers nor for the NO answers.

Furthermore, an ERP verification of the extreme categories C-300 and C+300 had been conducted. Figure 6 presents the ERPs at the CZ electrode for these categories. A paired t-test for CZ and PZ revealed a significant difference for the C+300 category (p<0.05, see table 7 in the Appendix). It is obvious that the C+300 category of the lottery condition is marked by a higher positivity at the centro-parietal electrodes (see also figure 9 in the Appendix).

Depending on the stimulus event of the colored frame presentation, the ERPs are not distinguishable, as seen in figure 7. Based on this identical pattern no further statistical test was deemed necessary.

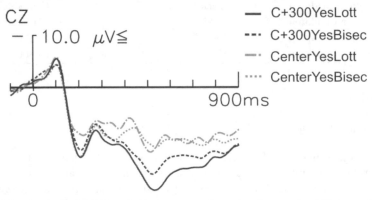

**Fig. 4.** Stimulus-locked ERPs of the inner number stimulus at the CZ electrode for the YES-answers

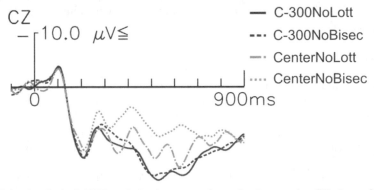

**Fig. 5.** Stimulus-locked ERPs of the inner number stimulus at the CZ electrode for the NO-answers

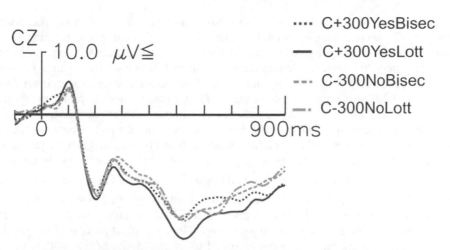

**Fig. 6.** Stimulus-locked ERPs of the extreme categories at the CZ electrode

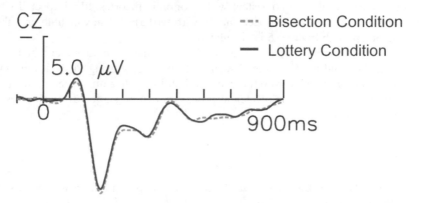

**Fig. 7.** ERPs of the colored frame stimulus

## 4  Conclusion

The experimental analysis shows three considerable results. First, there are no obvious differences between the bisection method and the CE method at the presentation of risk as well as for the behavioral data. Second, differences appear in the ERPs concerning the inner number categories. Third, the category C+300 in particular produces a high positive potential for the YES answers under the lottery condition.

The characteristic of the evoked ERPs within the time period of 450-600 ms post-stimulus shows that a P3 is existent. Given the assumption that a P3 reflects a task-related attention [13-16], the statement can be made that a higher positive potential represents higher attention on that stimulus.

The ANOVAs of the inner number stimulus revealed a very high significance for the time period of 450-600 ms, with the consequence that this stimulus will probably cause differences. ERPs of categories near the center position have a lower positive peak than categories further away from this center position, which is specifically category C+300 in the lottery condition. Thus, this stimulus induces a very high attention. Since in this category the sure payoff seems to be very attractive compared to the offered lottery, this potential could reflect the attractiveness of money or a pleasant surprise concerning the following sure payoff. Comparing this result with other EEG studies [20-22] analyzing the role of the P300 in connection with monetary reward, the argument that high attractive payoffs induce a higher P300 is reasonable.

The statistical analysis of the behavioral data shows no difference between the results of the CE method and the bisection method. It can be assumed that the perceived center of 'joy' equals the certainty equivalent, $CU \approx CE$ accordingly. Hence, the utility of both is $u(CU) = u(CE) = \frac{1}{2}$ and not $u(CU) > u(CE)$. There is no evidence that probabilities in risky choices are underweighted, as reported in some experimental or empirical papers on Prospect Theory. The EEG data shows no different attention on both methods at the presentation of risk. Thus, a distinct process of probability weighting or risk evaluation cannot be found for a probability of 0.5. Following the discussion in section 1.3 on modeling the evaluation of probabilities it can be concluded that these results are consistent with economic theories, like Expected Utility Theory or a special form of Prospect Theory with w(0.5)=0.5, in which no probability weighting of a probability of 0.5 is postulated.

Further experimental studies using EEG or fMRI techniques combining the bisection method and the CE method seem to be very promising to shed light on the phenomenon of probability weighting of probabilities that are different from 0.5. Economic models and models of mental processes can be subject to further tests by using these two methods.

# References

1. Kahneman, D., Tversky, A.: Prospect Theory: An Analysis of Decision under Risk. Econometrica 47(2), 263–291 (1979)
2. Tversky, A., Kahneman, D.: Advances in prospect theory: Cumulative representation of uncertainty. Journal of Risk and Uncertainty 5(4), 297–323 (1992)
3. Heldmann, M., Münte, T.F., Vogt, B.: Relevance without profit: Electrophysiological correlates of two economic methods for defining the utility function. Working Paper (2008)
4. Galanter, E.: The Direct Measurement of Utility and Subjective Probability. The American Journal of Psychology 75(2), 208–220 (1962)
5. Camerer, C.F., Ho, T.-H.: Violations of the betweenness axiom and nonlinearity in probability. Journal of Risk and Uncertainty 8(2), 167–196 (1994)
6. Tversky, A., Fox, C.: Weighting Risk and Uncertainty. Psychological review 102(2), 269–283 (1995)
7. Abdellaoui, M.: Parameter-Free Elicitation of Utility and Probability Weighting Functions. Management Science 46(11), 1497–1512 (2000)
8. Gonzalez, R., Wu, G.: On the Shape of the Probability Weighting Function. Cognitive Psychology 38(1), 129–166 (1999)
9. Loomes, G., Sugden, R.: Regret Theory: An Alternative Theory of Rational Choice Under Uncertainty. The Economic Journal 92(368), 805–824 (1982)
10. Bell, D.E.: Disappointment in Decision Making under Uncertainty. Operations Research 33(1), 1–27 (1985)

11. Albers, W., et al.: Experimental Evidence for Attractions to Chance. German Economic Review 1(2), 113–130 (2000)
12. Münte, T.F., et al.: Event-related brain potentials in the study of human cognition and neuropsychology. In: Handbook of neuropsychology, vol. 1, pp. 139–236 (2000)
13. Duncan-Johnson, C.C., Donchin, E.: On quantifying surprise: the variation of event-related potentials with subjective probability. Psychophysiology 14(5), 456–467 (1977)
14. Johnson, R.: The amplitude of the P300 component of the event-related potential: Review and synthesis. Advances in Psychophysiology 3, 69–137 (1988)
15. Linden, D.E.: The p300: where in the brain is it produced and what does it tell us? Neuroscientist 11(6), 563–576 (2005)
16. Polich, J.: Updating P300: an integrative theory of P3a and P3b. Clin Neurophysiol. 118(10), 2128–2148 (2007)
17. Picton, T.W.: The P300 Wave of the Human Event-Related Potential. Journal of Clinical Neurophysiology 9(4), 456 (1992)
18. Pritchard, W.S.: Psychophysiology of P300. Psychological Bulletin 89(3), 506–540 (1981)
19. Joyce, C.A., Gorodnitsky, I.F., Kutas, M.: Automatic removal of eye movement and blink artifacts from EEG data using blind component separation. Psychophysiology 41(2), 313–325 (2004)
20. Goldstein, R.Z., et al.: Compromised sensitivity to monetary reward in current cocaine users: an ERP study. Psychophysiology 45(5), 705–713 (2008)
21. Wu, Y., Zhou, X.: The P300 and reward valence, magnitude, and expectancy in outcome evaluation. Brain Res. (2009)
22. Yeung, N., Sanfey, A.G.: Independent coding of reward magnitude and valence in the human brain. J. Neurosci. 24(28), 6258–6264 (2004)

# Appendix

The Appendix can be found on:

http://www.ww.uni-magdeburg.de/pbwl1/07_experiments/
Evaluation_of_Probabilities_and_Brain_Activity_APPENDIX.pdf

# Human Factors Affecting Decision in Virtual Operator Reasoning

Lydie Edward, Domitile Lourdeaux, and Jean-Paul Barthès

Heudiasyc Laboratory UMR 6599 CNRS, University of Technology of Compiègne,
Centre de Recherches de Royallieu, 60200 Compiègne, France

**Abstract.** In this paper we present our research on virtual operator cognitive modeling and reasoning. They operate in a virtual environment for risk prevention. We want to model the influence of their cognitive states and personality on their decisions and actions in the environment. The purpose of our research is to design a system that generates behavior-based errors to support learning and risk prevention. It uses new mechanisms taking into account human factors and human behavior model regarding risky situations. We make use of the COCOM and CREAM model describing the different states or control mode in which an agent can be regarding to a lack of time or a temporal pressure. A challenge is to implement these models to produce the expected flexible, contextual and erroneous behaviors in both normal and constrained working conditions.

## 1 Introduction

The emergence of knowledge engineering tools for risk analysis and researches in cognitive and behavioral modeling give us the opportunity to develop tools to improve training and decision making for managing a team and/or preventing risks.

In our work, our goal is to highlight the link between some risky and damaged situations and the cognitive state of a human operator. We want to explain why in a certain context the operator choose a way than another. We want to introduce unexpected variations so that training scenarios are no longer predefined and ideal. This feature is particularly interesting when training objectives are not only to acquire the correct procedure, but also to highlight the difficulties of cooperative work and the constraints linked to hazardous professional activities. One of our particular concerns is also to understand how errors can be linked explicitly to situational, human and socio-organizational dimension. It is widely acknowledged that human and organizational factors play a critical role on human decisional process. The question is what is the most influential factors that affects decision ? Our system aims to take into account various situational constraints and human-factors issues to support the generation of the subsequent variations in the way virtual operators are behaving and applying the procedures to achieve their task. Such constraints could be the result of the combination of cognitive and physical parameters such as tiredness, stress, expertise, hunger,

N. Zhong et al. (Eds.): BI 2009, LNAI 5819, pp. 84–95, 2009.

motivation, excessive mental and physical workload. They can also come from time pressure or missing tools. We aim to emulate the deviated behaviors that could occur (errors, failure or botching) with a reasonable level of faithfulness. The scientific interests related to our work correspond to the following questions: how to represent in a faithful way the activity of a real operator in virtual environment? How do operators work in a deteriorated environment ? How do experts operate deviations and take some risks ?

In this paper we first present some models issued from cognitive studies designed for modeling errors and deviated human behaviors. Then we introduce our approach. After that, we present how we integrate these models into the decisional process of an operator and we finally present some results.

## 2   Errors and Behaviors Model

To understand the consequences of deviated behaviors on an organizational, human and technical system it is useful to introduce models that describe variable, contextual, erroneous and some time partial decisional process. Researches on human error produced various models and classifications to explain or predict errors occurrence in human behavior : decision scale of Rasmussen [10], human error mechanism [11], control mode and erroneous actions [4] [5] [8].

Rasmussen [10] classified human performance into three categories : Skill, Rule and Knowledge (SRK). *Skill-based behaviors* are routine activities conducted automatically and do not required conscious allocation of attentional resources. Behaviors are skill-based when human performance is determined by stored, preprogrammed patterns of instructions. *Rule-based behaviors* are activities controlled by a set of stored rules or procedures. The distinction between behaviors classified as skill-based and rule-based depends on the attention and training level of each individual. *Knowledge-based behaviors* are those in which stored rules no longer apply and a novel situation is presented for which a plan must be developed to solve a problem. In contrast to set rules, plans are often required to be changed based on the situation. Attentional resources must be allocated to the behavior and, therefore, the performance of knowledge-based behaviors is goal-controlled.

Reason [11], based on Rasmussen considerations, proposed a classification and differentiated three types of errors:

1. *Errors founded on automatism*, where actions deviate from followed intention, further to faults in execution (wool-gathering) or the stocking (memory), or else further to the application of an inadequate automatism (press on the bad key). We can perceive these errors only during the action execution.
2. *Errors founded on rules*, consist of a bad application of rules during the resolution of a problem (for example use the bad algorithm of resolution).
3. *Errors founded on knowledge*, consist of a bad use of knowledge while resolving a problem.

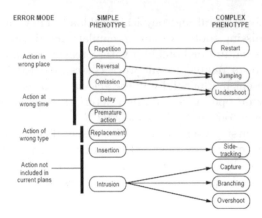

**Fig. 1.** CREAM model

In CREAM, Hollnagel [6] characterized the erroneous actions as a set of generic relation-ships between the human actions and what can be observed in terms of task goal achievement. Hollnagel calls *phenotype* the patterns of erroneous action that are observable and *genotype* the patterns in the underlying causal mechanisms that lead to phenotypes. These statements are based on genetic metaphors. The phenotypes are structured in simple (concerning one single action) and complex (covering complex sets of actions) phenotypes. The erroneous action are classified as follows (Fig. 1) :

1. *Repetition* : The previous action (or set of actions) is repeated.
2. *Inversion* : The order of two neighboring actions is reversed.
3. *Omission* : An action is not carried out. This can be inside a sequence of actions or - very often - the omission of the last action(s) of a series.
4. *Delay* : An action (or set of actions) started too late.
5. *Premature action* : An action (or set of actions) started too early, e.g., before a signal was given or the required conditions had been established.
6. *Replacement* : A simple task (action) is replaced by another simple task (action). Both are subtasks of a more general task.
7. *Insertion* : A simple task (action) belongs to the same general task is added in a sequence.
8. *Inclusion* : A simple task (action) is an inclusion when it is added in a sequence and has no link with the current task (not in the same hierarchy).

In the COCOM model, Hollnagel [5] defined four types of control mode corresponding to time zones within which an agent can operate :

1. *Strategic* : the agent has a broad time horizon and looks ahead at higher-level goals, having both an overall and a detailed view of the work system.
2. *Tactical* : this mode characterizes situations where performance more or less follows a known procedure or rule. The user's time horizon goes somewhat beyond the dominant needs of the present, but planning has a limited range.

The simplest situation will often be chosen, and safety constraints are therefore not always respected.

3. *Opportunistic* : the next action reflects the salient features of the current context. Only a little planning or anticipation is involved, perhaps because the context is not clearly understood by the agent or because the situation is chaotic. Opportunistic control is a heuristic that is applied when the knowledge mismatch is large, either due to inexperience, lack of formal knowledge, or an unusual state of the environment.

4. *Scrambled* : the next action is in practice unpredictable or random. Such a performance is typically the case when people act in panic, when cognition is effectively paralyzed and there is accordingly little or no correspondence between the situation and the actions.

## 3   Our Approach

Several researchers developed applications for modeling virtual operators reasoning [13] [3]. Only a few works propose to build systems incorporating behaviors based on cognitive and error models [2] [14] [15]. But this generation of behaviors is not founded on models issued from cognitive science, it is based on informatics foundation.

We find an application of the CREAM model in El Kechaì and Despres work [1]. They used it to know what errors the trainee is making and furthermore what causes lead to these errors. Woods [16] developed an application, based on Reason, reproducing a simplified functioning model of a central nuclear with a cognitive simulator to reproduce human operator actions. They do not aim to generate errors but to explain the errors done by a virtual operator controlled by a real human.

To characterize the deviations that virtual operators may sometimes display, we were inspired by the notion of borderline tolerated conditions of use (BTCU) from studies in ergonomics and human reliability [9]. This notion highlights the implicit individual and social regulations in the working environment which may lead to compromises in the use of tools and the performance of tasks. Some tasks are carried out only partially or not at all, because of a lack of time due to compromises made between safety and production. This concept is complementary to other elements linked to the individual, such as consciousness of risks, the effects of tiredness, or temporal pressures on performance.

In our approach, we combine COCOM, CREAM and BTCU. Starting from these models we propose new mechanisms and algorithms to simulate human decisional processes and generate human behavior-based errors. A challenge is to implement these models to produce the expected flexible, contextual and erroneous behaviors in both normal and constrained working conditions.

We integrate these models into a more global agent cognitive architecture, MASVERP[1], based on Belief-Desire-Intention [12] in order to produce behaviors in response to agent goals and perception of the environment. According to its

---

[1] Multi-Agent System for Virtual Environment for Risk Prevention.

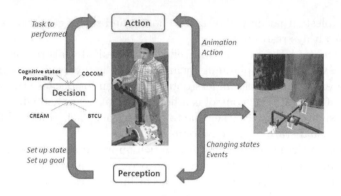

**Fig. 2.** Perception-Decision-Action loop

perception[2] (new events, leak, fire, an order, a changing state), the agent sets up its characteristics, its goal, its control mode and then the decisional process decides the next action (task) to perform. This decision generate an action in the environment which can succeed or not. Fig. 2 illustrates this mechanism.

## 4   Decisional Process

### 4.1   Task Representation

To represent agents activity we use HAWAI-DL[3] language that consists of a formal description of the task as an arborescent decomposition beginning from the more abstract task (root) decompose with subtasks until the lower level, the action (leaf). The description of the activity integrates individual features and situational constraints. A task is constitute with three parts : the core, relations with the world and the link with the objects.

The core corresponds to all the information needed to describe the task. Relations with the world are the world states managing the execution of the task (starting conditions, realization or stop conditions).

The scheduling of the subtasks of a task is defined by different types of constructors: SEQ (subtasks are executed in the given order), ALT (only one subtask is executed), AND (subtasks are executed in any order and do not share data and resources), OR (subtasks are executed in any order and at least one subtask is executed), PAR (subtasks are executed in any order and share data and resources), SIM (subtasks are executed at the same time).

The context of realization of the task is described with preconditions attributes. The consequences of the task execution is describes with post-conditions attributes. For the preconditions we distinguish :

---

[2] Perception is the ability to take in information via the senses, and process it in some way.

[3] Human Activity and Work Analysis for sImulation-Description Language.

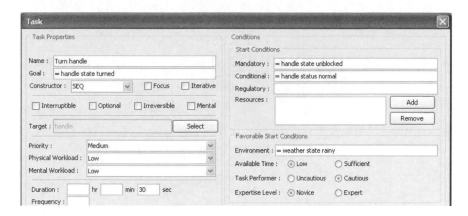

**Fig. 3.** A sample of a task described with HAWAI-DL

1. *Mandatory* : state of the world required to perform the task
2. *Regulatory* : state of the world required by safety regulations
3. *Resources* : tools that are required to accomplish the task
4. *Safety* : Safety tasks are compromise between safety and production. They are usually done only by expert.
5. *Favorable* : the context in which the task is relevant, for example *favorable expertise condition*. There is two ways to remove a pipe. The expertise condition can be: bolts are not rusted. If the bolts are rusted, an expert will do the task with the good expertise conditions.

A task has conditional and dynamic states. When a precondition is true the task state is set to active for this precondition: mandatory active, regulatory active, favorable active, resource active. These are the conditional states. The dynamic states are used during plan execution : active/inactive, pending/finished, conflicting, concurrent, failed.

Preconditions as well as postconditions have the same formalism. Both are conditions. Conditions can be combined with some logical operators (AND, OR). We represent conditions as a quadruple: $< operator, object, property, value >$.

Example : $(= Pipe_6\ status\ normal), (> Pipe_6\ diameter\ 50)$.

### 4.2   Reasoning with COCOM, BTCU and CREAM

According to the COCOM model, the parameter that influences the decision and behavior of an agent is the temporal pressure. We added some supplementary parameters that influence according to us operators control mode and consequently their behaviors. We integrate : (i) the internal state of an agent, (ii) temporal pressure which comes from the agent itself, (iii) equipment, (iv) environmental parameters, (v) parameters of task (Fig. 4). We distinguish three categories of internal states: cognitive (risk perception, cognitive workload, vigilance, motivation), physical (strength, physical workload) and physiological (stress, hunger, thirst, tiredness, agitation).

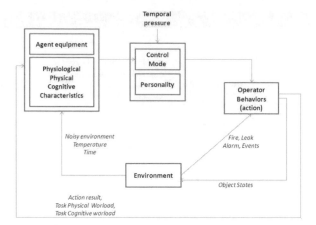

**Fig. 4.** Link between main influential factors on Control Mode

|  | Temporal Pressure | Anticipation | Planning/performance |
|---|---|---|---|
| **Strategic** | None | Large | • Global context and higher goals are considered<br>• Efficient and robust performance<br>• Tools pre-planning |
| **Tactic** | Acceptable | Middle | • Performance based on limited scope of planning<br>• Follows a known procedure or rule<br>• Rules are not respected<br>• Tools pre-planning |
| **Opportunistic** | Middle | Bad | • Very little planning, actions are often based on perceptually dominant or most frequently used features<br>• Context is not clearly understood<br>• Analogy planning , Focus<br>• Tools pre-planning |
| **Scrambled** | High | None | • Next action is unpredictable , No planning<br>• Little or no thinking involved<br>• Complete loss of situation awareness |

**Fig. 5.** Planning depending the control mode

We use the control mode ($C_m$) as follows and as describes in Fig. 5. In strategic, tactical and opportunistic mode agent will prepare the intervention (tools). Depending on the mode, operators are doing the good procedures or operate deviations or even errors or violations. They do not interpret the context in the same way and do not have the same representations.

We integrate safety implications directly in the description of a task. Safety tasks (BTCU tasks) may be characterized as "'allowed to be performed"', "'allowed not to be performed"', or "'not allowed"'. According to their $C_m$ agents will do or not the BTCU tasks (Fig. 6).

To determine the errors that can be done according to CREAM, we consider CREAM model as a request base. In this article we only consider the simple phenotype proposed by Hollnagel (Fig. 7). We determined which characteristics, parameters can bring the agent to do such errors and associated it with rules. We also linked errors with the $C_m$.

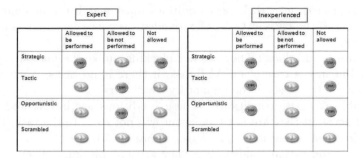

**Fig. 6.** Does agent can apply a BTCU task ?

| Simple phenotype | Mode | Influential Parameters |
|---|---|---|
| Repetition | Tactic<br>Opportunistic | Inattention, Memory failure<br>Cautious (very), Unexperienced |
| Reversal | Scrambled | Stress |
| Omission | Opportunistic<br>Scrambled | Focus Task, Memory failure<br>Lack of Knowledge |
| Delay | All mode | External events,<br>Noisy environment<br>Collaborative work<br>Tiredness, Lack of attention<br>Communication failure |
| Premature action | Scrambled<br>Opportunistic | Stress, Hurried, Force |
| Replacement | Scrambled<br>Opportunistic | Lack of Knowledge<br>Imprudent, Inattention |
| Insertion | All mode | External events |
| Inclusion | Opportunistic<br>Scrambled | Hurried<br>Stress |

**Fig. 7.** Dependencies between error, control mode and characteristics

## 4.3   How Does Agent Interpret the Task Conditions ?

In the interpretation of the preconditions we add rules linked to the cognitive
state of the agent (example : favorable expertise conditions are evaluated only by
expert agent, strategic agent does not performed safety related task that is not
allowed). Preconditions are tested typically by checking the status (value) of the
concerned object. This can be done by sending a query to the object database.
We describe below how each preconditions are evaluated during the execution
and Fig. 8 shows the implemented function.

1. Mandatory conditions : If they are not true, the task can not be executed.
   Agents are trying to make them true by replanning. The state of the current
   task is set to pending. The process looks for tasks that responding to the

following constraint : their postconditions must match the mandatory pre-conditions of the current task. The founded tasks are executed as a general OR-task.

2. Regulatory conditions : They are not blocking unlike the mandatory conditions. But in the case of an ALT-task, if there is no tasks to execute, the planner can try to solve the regulatory conditions.

3. Favorable expertise conditions : These conditions are interpreted only by expert agent. If they are not true agent will not execute the task. Example of environment conditions : $(>$ $Pipe_6$ $diameter$ 50)

4. Favorable contextual conditions : They are not blocking. They are useful in the case of an OR-task, if all the tasks can be executed then the conditions allows to choose the more appropriate task depending on the context. Example of contextual conditions : $(=$ $weather$ $state$ $rainy)$

5. Safety conditions : Same as the favorable conditions, only expert agent will evaluate the safety environment conditions.

6. Resources conditions : First we check if agent has the resource in its tool box, in its pockets or in its hands. If agent does not have the resource, if it is not in a hurry it will look for a plan to get the resource. It has also the possibility to ask another agent. If agent is in opportunistic mode it will reason by analogy. This means it looks for a similar tool that can be used to do the task (a-sort-of object).

## 5   Results

We developed each operator as a cognitive autonomous agent on the OMAS platform. We have an agent in charge of managing the objects database and answering queries. Each operator agent has an interface agent called MIT[4] which is in charge of sending to its interface module developed with QT all information about the agent's states, the value of its different characteristics and its plan. These values are use to show the evolution in real time of the agent's states and plans. We tested our algorithms on a small part of the scenario (Fig. 9).

Agents have to do the task "Drain the pipe". This order matches the goal $g_1(=$ $pipe_06$ $state$ $drained)$. The planner generates a plan to achieve this goal. Fig. 10 illustrates the action selection mechanism results for three steps. Yellow tasks are $active$, greens are $ended$ and red are $in$ $failure$. At step 1, $Agent_1$ chose to prepare the recuperation by putting a bucket under the $Pipe_06$. The other choice is a task related to safety, inexperienced agent can not do this type of task. $Agent_2$ that is expert, chose the second way that is to throw absorbent and ammonia on the floor. The second main task of this little scenario is to open the gate so that the liquid can flow. There is one of two ways to do the task : turn the handle of the gate slowly or roughly. $Agent_1$ chose the first alternative that is to turn slowly. As the handle is rusted the task

---

[4] Masverp InTerface.

```
(if (and (not (mv-activable-mandatory? task))
         (has-v-mandatory task))
    (solve-mandatory agent (car (has-v-mandatory task)) task))
(if (mv-task-verify-mandatory? task)
    (send task '=replace "v-activable-mandatory" 't))

(if (and (not (mv-activable-regulatory? task))
         (has-v-regulatory task))
    (if (or
            (and (agent-expert?)
                 (agent-strategic?))
            (and (agent-inexperienced?)
                 (or (agent-strategic?)
                     (agent-tactic?))))
        (progn
            (solve-regulatory agent (car (has-v-regulatory task)) task)
            (if (mv-task-verify-regulatory? task)
                (send task '=replace "v-activable-regulatory" 't)))))

(if (and (has-v-expertise task)
         (agent-expert))
    (if (eval (car has-v-expertise))
        (send task '=replace "v-activable-expertise" 't)
      (send task '=replace "v-activable-expertise" 'nil)))

(if (has-v-contextual task)
    (if (or
            (and (or (agent-expert?) (agent-inexperienced?))
                 (or (agent-strategic?) (agent-tactic?))))
        (if (eval (car has-v-contextual))
            (send task '=replace "v-activable-contextual" 't)
          (send task '=replace "v-activable-contextual" 'nil))))

(if (and (car (has-v-tclu task))
         (applicable-clu? task agent))
    (progn
      (send task '=replace "v-activable-tclu" 't)
      (send task '=replace "v-activable-tclu" 'nil)))

(if (or (null (mv-task-resource? task))
        (contains-tools resource (omas::recall masverp :tools)))
    (send task '=replace "v-activable-resource" 't))
```

**Fig. 8.** Check Conditions Function

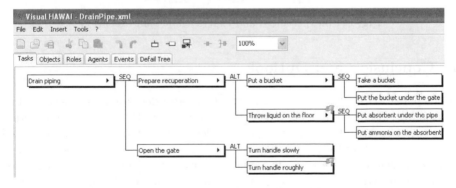

**Fig. 9.** Extract of the task model

does not succeed. In the remainder of the simulation, agent looks for a task to achieve the goal $g_2(= handle_05 \; state \; unblock)$. $Agent_2$ chose the same task but it also tries the other alternative related to safety that is to turn roughly the handle. It does not succeed. It will also look for a way to unblock the handle.

| | T | State of the world | Agent state | CM | Actions | Action Selection |
|---|---|---|---|---|---|---|
| A1 | $\tau_1$ | = handle05 state rusted | Inexperienced | T | Take bucket05 | 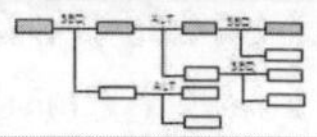 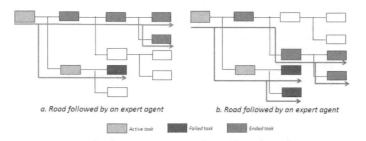 |
| A2 | $\tau_1$ | = handle05 state rusted | Expert | T | Put Absorbent2 under Pipe06 | 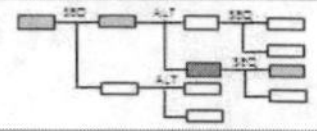 |
| A1 | $\tau_2$ | = handle05 state rusted | Inexperienced | T | Put Bucket05 under Pipe06 | 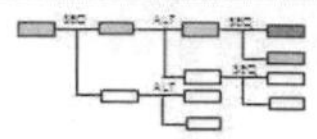 |
| A2 | $\tau_2$ | = handle05 state rusted | Expert | T | Put Ammonia2 on Absorbent2 | 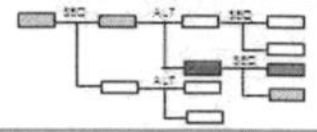 |
| A1 | $\tau_3$ | = handle05 state rusted | Inexperienced | T | Turn handlle05 | 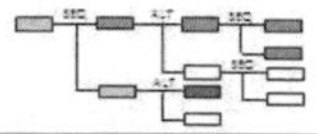 |
| A2 | $\tau_3$ | = handle05 state rusted | Expert | T | Turn handle05 | 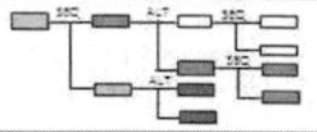 |

**Fig. 10.** Planning Execution

a. Road followed by an expert agent    b. Road followed by an expert agent

Active task    Failed task    Ended task

**Fig. 11.** Inexperienced and expert agent plan execution

## 6   Conclusion and Outlook

In this paper we present our work on virtual operator decisional process modeling using models issued from cognitive science and safety. By introducing different categories of preconditions and cognitive rules we are able to generate deviated behaviors according to agent personality and competencies. An expert or inexperienced agent does not have the same reasoning, knowledge and representation and consequently their actions are not always the same. We are able to trace agent's activity and explain the observed behaviors according to agent's personality, physiological, physical and cognitive states. We are also able to trace the evolution of these parameters.

The next step of our work will be to enrich the decisional module by integrating other types of errors according to CREAM model and to test if our model still behaves in an acceptable way.

# References

[1] El-Kechai, N., Despres, C.: A Plan Recognition Process, Based on a Task Model, for Detecting Learner's Erroneous Actions. In: Intelligent Tutoring Systems, pp. 329–338 (2006)

[2] Funge, J., Tu, X., Terzopoulos, D.: Cognitive Modeling: Knowledge, Reasoning and Planning for Intelligent Characters. In: Proc. of SIGGRAPH 1999, Los Angeles, CA (1999)

[3] Le Strugeon, E.G., Hanon, D., Mandiau, R.: Behavioral Self-control of Agent-Based Virtual Pedestrians. In: Ramos, F.F., Larios Rosillo, V., Unger, H. (eds.) ISSADS 2005. LNCS, vol. 3563, pp. 529–537. Springer, Heidelberg (2005)

[4] Hollnagel, E.: The phenotype of erroneous action. International Journal of Man-Machine Studies 39, 1–32 (1993)

[5] Hollnagel, E.: Human Reliability Analysis: Context and Control. Academic Press, London (1993)

[6] Hollnagel, E.: Cognitive Reliability and Error Analysis Method - CREAM. Elsevier, Amsterdam (1998)

[7] Hollnagel, E.: Time and time again. Theoretical issues in Ergonomics Science, 143–158 (2002)

[8] Hollnagel, E.: Barriers and accident prevention. Ashgate, Aldershot (2004)

[9] Polet, P., Vanderhaegen, F., Amalberti, R.: Modelling border-line tolerated conditions of use (BTCU) and associated risks. Safety Science journal, 111–136 (2003)

[10] Rasmussen, J.: Skills, rules, and knowledge; Signals, signs, and symbols, and other distinctions in human performance models. IEEE transactions on systems, man, and cybernetics, SMC 13, 257–266 (1983)

[11] Reason, J.: Human error. Cambridge University Press, Cambridge

[12] Rao, A., Georgeff, M.: BDI Agents: From Theory to Practice. In: Proc. 1st Int. Conf. on Multi-Agent Systems (ICMAS 1995), San Francisco, CA, pp. 312–319 (1995)

[13] Rickel, J., Johnson, L.: Extending virtual humans to support team training in virtual reality. In: Lakemayer, G., Nebel, B. (eds.) Exploring Artificial Intelligence in the New Millenium, pp. 217–238 (2002)

[14] Shao, W., Terzopoulos, D.: Autonomous Pedestrians. Graphical Models 69(5-6), 246–274 (2007)

[15] Thomas, G., Donikian, S.: Virtual humans animation in informed urban environments. In: Computer Animation, pp. 129–136. IEEE Computer Society Press, Philadelphia (2000)

[16] Woods, D., Roth, E.: Cognitive Engineering: Human Problem Solving with Tools. Human Factors 30(4), 415–430 (1988)

# The Effect of Information Forms and Floating Advertisements for Visual Search on Web Pages: An Eye-Tracking Study

Mi Li[1,2], Jingjing Yin[1], Shengfu Lu[1], and Ning Zhong[1,3]

[1] International WIC Institute, Beijing University of Technology
Beijing 100022, P.R. China
lusf@bjut.edu.cn
[2] Liaoning ShiHua University, Liaoning, 113001, P.R. China
limi_666@emails.bjut.edu.cn
[3] Dept. of Life Science and Informatics, Maebashi Institute of Technology
Maebashi-City 371-0816, Japan
zhong@maebashi-it.ac.jp

**Abstract.** Users' visual search on a Web page is impacted by information forms, information layout, Internet advertisements (ads for short), etc. Text and picture are two important forms of expressing the information on Web pages, and it is generally through the two forms of title that users can search their desired information. This study investigates the effect of the two basic information forms and floating ads on visual search using eye-tracking. By analyzing the visual search time and pupil diameter, the results show that it is easier to find the picture than the text; whether the target is text or picture, floating ads do not significantly impact people's visual search time, however, it would make people bored.

## 1 Introduction

A website includes a series of related Web pages, and each Web page contains various forms of information. In general, the title of information on a Web page is expressed as the form of text or picture, and it is through the two forms of title that users search their desired information. However, the users' visual search on Web pages is different from the search engines like Google, which will be impacted by many factors [1] - [4], such as information layout, information forms, and Internet advertisements.

The study about text and picture reported that picture had the superiority effect to text [5]. Carroll et al. [6] through the eye movement study also found that users would look at the picture information first, and then read the text. Brandt also indicated that majority of users paid more attention to the picture, whereas the text was read at the end in his eye tracking studies [7]. Rayner et al. [8] reported that when participants looked at the print ads, the average fixation duration in picture of ads was significantly longer than text, which also provided the evidence that people pay more attention to picture than text. However, these

N. Zhong et al. (Eds.): BI 2009, LNAI 5819, pp. 96–105, 2009.

studies about text and picture which are just simple visual stimuli are different from Web pages which have a variety of influencing factors.

There are also many studies about the effect of Internet ads on the user's visual browsing behavior, such as ads content [9], present form [10] and location [11] which had the effect on banner ads. Stenfors et al. through exploring internet ads found that the Internet ads had no significant effect for users' visual search on Web pages [12,13]. Balkenius and Moren found that users wouldn't pay attention to the useless Internet ads when they were browsing on Web pages [14]. These studies indicate that the most users will neglect the existence of Internet ads during visual browsing information on Web pages. However, a study demonstrated that animation can disperse users' attention to the content of Web pages [15]. Based on our knowledge, it has not been reported about the study on the effect of users' visual search for the usual condition that the Web page with a floating ad.

An eye tracker is a mental measure precision instrument which can capture the changes of eye movement and record the data, such as pupil size and scan path during people reading, looking at a picture or viewing scenery [16].

In this study, participants' eye movement data were recorded by Tobii T120 eye tracker. By analyzing the visual search time and pupil diameter, we investigated the two basic information forms and floating ads for the effect of the visual search behavior on Web pages.

## 2   Experimental Design and Method

### 2.1   Experimental Web Pages

In the study, we designed 10 Web pages matching 10 search targets, which were involving different topics such as the mobile, car and diet. The searched target of each Web page was either text or picture, text was the Chinese phrase consisted of $3 \sim 6$ Chinese words, and the picture was a well-known logo picture which was almost the same as text in visual effects. We arranged the searched targets on different locations to eliminate the effects of information layout for the visual search time, as well as used the single page type with no scrollbar to eliminate the effects of using scrollbar for visual search time.

The experiment included visual search on the Web page without floating ads and with floating ads. The Web page with floating ads was just adding a floating ad on the Web page without floating ads. The floating ad was displayed in synchronism with the Web page, which appeared at random position and floated as the same speed as the real situation on the Web page. Fig. 1 shows the possible pathways of the floating ad on a Web page.

### 2.2   Participants and Apparatus

The participants were 100 undergraduates or postgraduates from various majors with the average age of 23.0 years old $(SD = 1.8)$, in which 51 were female. All participants were right-handed, skilled users of the Internet, had normal

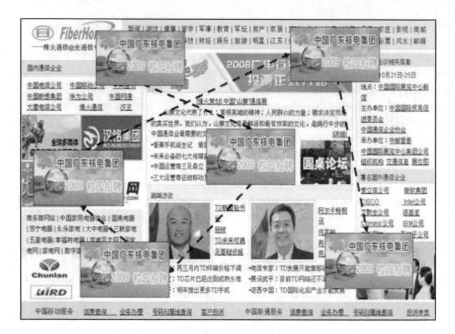

**Fig. 1.** Possible pathway of the floating ad on a Web page. The dotted lines represent the possible pathway of the floating ad.

or corrected-to-normal vision, and none of them had the experience about the eye movement experiment. Participants were randomly divided into 2 groups according to whether the Web page with floating ads. There were 50 participants in the group without floating ads, and the other 50 participants were in the group with floating ads. Eye tracker recorded the eye movement data in the whole visual search process from participants' viewing the Web page to finding out the required phrase or logo picture.

We used Tobii T120 eye tracker to record participants' eye movement during them either searching the text or logo picture on Web pages without floating ads and with floating ads. The rate of Tobii T120 eye tracker is 120 HZ, the Web pages were displayed on the screen which is a 19" LCD monitor with a resolution set to 1024 × 768 pixels.

## 3   Results

In this study, we focus our analysis on the search time and pupil diameter during participants searched text and logo picture on the Web page with and without floating ads.

## 3.1    Search Time

**Search Time on Text vs. on Logo Picture.** We contrasted visual search time for searching both text and logo picture on the Web page with floating ads or without floating ads. As shown in Fig. 2(a), the average search time on text (14.16 s) was longer than on logo picture (8.69 s) and there was a significant difference between them when participants performed visual search on the Web page without floating ads $[F(1, 98) = 40.83, P < 0.05]$. As shown in Fig. 2(b), the average search time on text (14.91 s) was also longer than on logo picture (7.90 s) and there was a significant difference between them when participants performed visual search on the Web pages with floating ads $[F(1, 98) = 71.34, P < 0.05]$. The results indicated that, whether there were floating ads on Web pages or not, the average search time on text was significant longer than on logo picture.

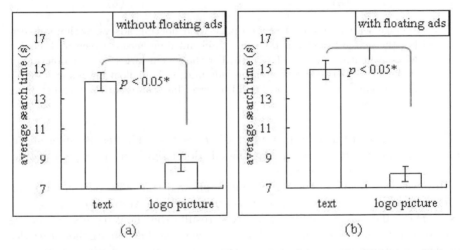

**Fig. 2.** Average search time on text vs. on picture. (a) For the Web page without floating ads, the average search time on text was significant longer than on picture (p < 0.05*). (b) For the Web page with floating ads, the average search time on text was significant longer on picture (p < 0.05*). 0.05 is the significant level of F-check, and the error lines represent the mean standard error.

**Search Time on the Web Page with Floating Ads vs. without Floating Ads.** We contrasted visual search time for searching either text or logo picture on the Web page with floating ads and without floating ads, as shown in Fig. 3(a). Although the average search time on the Web page with floating ads (14.91 s) was longer than without floating ads (14.16 s) when the search target was presented as text, the difference between them was not significant $[F(1, 98) = 0.68, P > 0.05]$. As shown in Fig. 3(b), even though the average search time on the Web page with floating ads (7.90 s) was shorter than without floating ads (8.69 s) when the search target was presented as picture, there were no significant

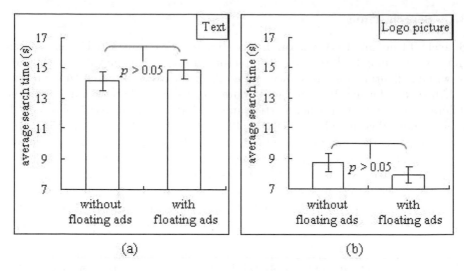

**Fig. 3.** Average search time on the Web page with floating ads vs. without floating ads. (a) For the search target was text, no significant difference between with floating ads and without floating ads (p > 0.05). (b) For the search target was logo picture, no significant difference between with floating ads and without floating ads (p > 0.05). 0.05 is the significant level of F-check, and the error lines represent the mean standard error.

differences between them [$F(1, 98) = 1.02, P > 0.05$]. The results suggest that no matter search text or picture, floating ads don't significantly impact people's visual search time on Web pages.

However, the result of the questionnaire showed that 83% participants considered floating ads impacting their visual search on Web pages [$\chi^2_{0.05}(1) = 43.56, P < 0.05$], which conflicted with the results that floating ads had no significant effects on the search time. Therefore, we would do the further statistical analysis of pupil diameter which is related to emotion in cognitive science.

## 3.2  Pupil Diameter on the Web Page with Floating Ads vs. without Floating Ads

Figure 4 presents the changes in pupil diameter when participants did visual search on the Web page with or without floating ads. For the search target was presented as text, the average pupil diameter on the Web page with floating ads (3.58 mm) was smaller than without floating ads (3.79 mm) and there was a significant difference between them [$F(1, 98) = 5.60, P < 0.05$], as shown in Fig. 4(a). For the search target was presented as logo picture, the average pupil diameter on the Web page with floating ads (3.62 mm) was also significant smaller than without floating ads (3.84 mm)[$F(1, 98) = 6.67, P < 0.05$], as shown in Fig. 4(b). A study demonstrated that the changes in pupil diameter reflect the changes in emotion: the pupil dilated in pleasant emotion and pupil constricted in annoying emotion [17]. The result of the pupil diameter indicates

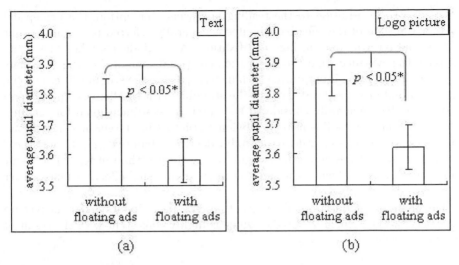

**Fig. 4.** Average pupil diameter in the Web page with floating ads vs. without floating ads. (a) For the target is text, with floating ads is significant smaller than without floating ads ($p < 0.05$*). (b) For the target is logo picture, with floating ads is significant smaller than without floating ads ($p < 0.05$*). 0.05 is the significant level of F-check, and the error lines represent the mean standard error.

that no matter the search target presents as text or picture, the floating ads impact users' mood that it would make people bored when they perform visual search on Web pages.

## 4    Discussion

### 4.1    Information Form Impacts Users' Visual Search Efficiency

Search time is the main indicator for measuring users' visual search efficiency on Web pages. The shorter the visual search time is, the higher the visual search efficiency would be. This study investigated the effect of information forms including text and logo picture for users' visual search on Web pages. Results indicate that the information form significantly impacts the visual search efficiency on Web pages, and whether there are floating ads on Web pages or not, the visual search time on text is significantly longer than logo picture (see Fig. 2).

Text on Web pages is often in a single color, not colorful as the picture. Therefore, several previous studies show that the colorful pictures attract users' more attention than text. Users usually view the picture before reading the text [6], which because the color of the picture could provide prior clues for their visual search and attract their eyeballs [18,19]. Rayner also pointed out that viewers' initial fixation frequency on the picture (69%) are much higher than the text (16%) [20]. Moreover, When a hyperlink is presented as two forms text and picture on the same page, users often prefer clicking the picture to the text [21]. In addition to the reading and scene perception, the perceptual span or the span of the effective stimulus is generally referred to as the amount of information available during each fixation. A great deal studies [22] - [26] supplied the evidences that the larger the perceptual span of the stimulus refers the more information input during each fixation. Thus, the stimulus is easy to be identified and processed, in contrast, the stimulus with smaller perceptual span is difficult. It is generally considered that text is one-dimensional for reading, whereas picture is two-dimensional. Obviously, picture has larger perception span than text, which makes picture easier to be identified than text [27]. In our study, the visual search time in picture is significantly shorter than in text. The result shows that picture which is the same as the text in visual effects is much easier to be searched than text.

The result of the questionnaire shows that 84% participants prefer the picture to text $[\chi_{0.05}^2(1) = 46.24, P < 0.05]$, which is highly consistent with the result of the search time.

## 4.2   Floating Ads Impact on Visual Information Search

**Floating Ads don't Impact Visual Information Search Efficiency.** As one of the most popular Internet ads, floating ads usually appear on the Web page when users view a new Web page. In this study, we further investigated the effect of floating ads for visual search efficiency on visual searching the text or picture.

User's visual behavior on Web pages includes visual browsing and visual search. Some studies about visual browsing on Web pages indicated that users wouldn't pay attention to the useless Internet ads [14]. In addition, the study on static banner ads also demonstrates that users usually take the avoidance strategies according to their experience [28], namely, most users often neglect the existence of Internet ads when they are browsing information on Web pages. A study about visual search on Web pages showed that the animation flash pictures would disperse the user's attention [15]. That is to say, animation ads may affect users' visual search time. However, there were also some studies showed that Internet ads had little effect for users' visual search behavior on Web pages [12,13].

The results of this study demonstrated that, whether the search target was text or picture, floating ads had no significant effect on users' search time (see Fig. 3).

The floating ads as a rich media type of Internet ads are usually involving the commercial activity, which are extension of the general media ads, such as paper ads and TV ads. Different from the general media ads, floating ads appear over the Web page, and people can selectively avoid or ignore them. Users often believe that floating ads are not useful for their requirement. Especially during users' searching information by eyes, they will be unconscious to consider that floating ads have no relation to their desired information, and adopt the avoidance strategies. In our study, whether the search target is text or picture, few participants looked at the floating ads during they performed the visual search tasks. Most participants adopted the avoidance strategies so that the search efficiency wasn't significantly impacted by floating ads. In reflection to the search time, the result shows that there is no significant difference on the search time between with floating ads and without floating ads.

### 4.3   Floating Ads Impact People's Mood

Previous researches [29,30] about the relationship between the pupil size and mental activity suggested that the pupil diameter dilated in positive emotion (such as happy and joy), and the pupil diameter constricted in negative emotion (such as fear and sadness). These results show that the pupillary response is associated with the changes in emotion.

By analyzing the average pupil diameter, we found that whether the searched target is text or picture, the average pupil diameter on the Web page with floating ads was significantly smaller than without floating ads. The result shows that users don't prefer searching on the Web page with floating ads, which makes the pupil diameters constricted. Therefore, floating ads would make people bored when they perform visual search on Web pages.

The result of questionnaire shows that 94% participants didn't like floating ads during them searching information on Web pages by eyes, which is highly consistent with the result of the average pupil diameter.

## 5   Conclusion

This paper reports an experiment study on the effect of information forms and floating ads for visual search using eye-tracking. The results show that the visual search time on text is longer, in contrast that on picture is shorter. And there is significant difference between them. Therefore, the picture which is the same as the text in visual effects is easier to be searched than text. In addition, the study also investigates the effect of floating ads. The results show that whether the searched target is text or picture, floating ads do not significantly impact people's visual search time, however, the result of the pupil diameter indicates that floating ads would make people bored.

## Acknowledgements

This work is partially supported by the National Science Foundation of China (No. 60775039 and No. 60673015) and the grant-in-aid for scientific research (No. 18300053) from the Japanese Ministry of Education, Culture, Sport, Science and Technology, and the Open Foundation of Key Laboratory of Multimedia and Intelligent Software Technology (Beijing University of Technology) Beijing.

# References

1. Russell, M.C.: Hotspots and Hyperlinks: Using Eye-tracking to Supplement Usability Testing. Usability News 72-Russell 2(7), 1–11 (2005)
2. Guan, Z.W., Cutrell, E.: An Eye Tracking Study of the Effect of Target Rank on Web Search. In: Proceedings of the SIGCHI Conference on Human Factors in Computing Systems, pp. 417–420 (2007)
3. Halverson, T., Hornof, A.J.: Local Density Guides Visual Search: Sparse Groups are First and Faster. In: Meeting of the Human Factors and Ergonomics Society, pp. 1860–1864. HFES Press, New Orleans (2004)
4. Weller, D.: The Effects of Contrast and Density on Visual Web Search, http://psychology.wichita.edu/newsurl/usabilitynews/62/density.asp
5. Maldonado, C.A., Resniek, M.L.: Do Common User Interface Design Patterns Improve Navigation? In: Proceedings of the Human Factors and Ergonomics Society 46th Annual Meeting, pp. 1315–1319 (2002)
6. Carroll, P.J., Young, J.R., Guertin, M.S.: Visual Analysis of Cartoons: A View from the Far Side. In: Rayner, K. (ed.) Eye Movement and Visual Cognition: Scene Perception and Reading, pp. 444–461 (1992)
7. Brandt, H.: The Psychology of Seeing. The PhilosoPhical Library, New york (1954)
8. Rayner, K., Rotello, C.M., Steward, A.J., et al.: Integrating Text & Pictorial Information: Eye Movements When Looking at Print Advertisements. Journal of Experimental Psychology: Applied 7(3), 219–226 (2001)
9. Briggs, R., Hollis, N.: Advertisement on the Web: Is there Response Before Click-through? Journal of Advertising Research 2(37), 33–45 (1997)
10. Faraday, P.: Attending to Web Pages. In: CHI 2001 Extended Abstracts on Human Factors in Computing Systems, pp. 159–160. ACM Press, WA (2001)
11. Benway, J.P.: Banner Blindness: The Irony of Attention Grabbing on the World Wide Web. In: Proceeding of the Human Factors and Ergonomics Society 42nd Annual Meeting, pp. 463–467. HFES Press, Chicago (1998)
12. Stenfors, I., Holmqvist, K.: The Strategic Control of Gaze Direction when Avoiding Internet Ads. In: 10th European Conference on Eye Movements, pp. 23–25 (September 1999)
13. Stenfors, I., Moren, J., Balkenius, C.: Avoidance of Internet Ads in A Visual Search Task. In: Eleven European Conference on Eye-Movements, pp. 22–25 (2001)
14. Balkenius, C., Moren, J.: A Computational Model of Context Processing. In: Proceedings of the 6th International Conference on the Simulation of Adaptive Behavior, pp. 256–265. The MIT Press, Cambridge (2000)
15. Zhang, P.: The Effects of Animation on Information Seeking Performance on the World Wide Web: Securing Attention or Interfering with Primary Tasks? Journal of the Association for Information Systems 1(1), 1–28 (2000)
16. Rayner, K.: Eye Movements in Reading and Information Processing: 20 Years of Research. Psychological Bulletin 124, 372–422 (1998)
17. Bellenkes, A.H., Wickens, C.D., Kramer, A.F.: Visual Scanning and Pilot Expertise: The Role of Attentional Flexibility and Mental Model Development. Aviation, Space, and Environmental Medicine 68, 569–579 (1997)
18. Triesman, A.: Features and Objects: Fourteenth Bartlett Memorial Lecture. Quarterly Journal of Experimental Psychology 2(40A), 201–237 (1988)
19. Fleming, J.: Defense of Web Graphics: Graphic Designers Offer More than Just Flashy Graphics, http://webreview.com/97/07/25/feature/index4.html

20. Rayner, K., Miller, B., Caren, M.R.: Eye Movements When Looking at Print Dvertisements: The Goal of the Viewer Matters. Applied Cognitive Psychology Appl. Cognit. Psychol. 22, 697–707 (2008)
21. Stenfors, I., Moren, J., Balkenius, C.: Behavioural Strategies in Web Interaction: A View from Eye-movement Research. Elsevier Science BV, 633–644 (2003)
22. Rayner, K.: The Perceptual Span and Peripheral Cues in Reading. Cognitive Psychology 7, 65–81 (1975)
23. Rayner, K.: Eye Movements in Reading and Information Processing. Psychological Bulletin 85, 618–660 (1978)
24. Rayner, K.: Eye Movements and the Perceptual Span in Beginning and Skilled Readers. Journal of Experimental Child Psychology 41, 211–236 (1986)
25. Saida, S., Ikeda, M.: Useful Field Size for Pattern Perception. Perception & Psychophysics 25, 119–125 (1979)
26. Van Diepen, P.M.J., Wampers, M.: Scene Exploration with Fourier-filtered Peripheral Information. Perception 27, 1141–1151 (1998)
27. Nelson, D.L., Castano, D.: Mental Representations for Picture and Words: Same and Different. American Journal of Psychology 97, 1–15 (1984)
28. Benway, J.P., Lane, D.M.: Banner Blindness: Web Searchers Often Miss "obvious" Links. Internet Working (2000),
    http://www.internettg.org/newsletter/dec98/banner-blindness.html
29. Hess, P.J.M.: Pupil Size as Related to the Interest Value of Visual Stimuli. Science 132, 349–350 (1960)
30. Hess: Attitude and Pupil size. Scientific American, 212, 46–54 (1965)

# Figural Effects in Syllogistic Reasoning with Evaluation Paradigm: An Eye-Movement Study

Xiuqin Jia[1], Shengfu Lu[1], Ning Zhong[1,2], and Yiyu Yao[1,3]

[1] International WIC Institute, Beijing University of Technology
Beijing 100022, P.R. China
xq.jia@emails.bjut.edu.cn, lusf@bjut.edu.cn
[2] Dept. of Life Science and Informatics, Maebashi Institute of Technology
Maebashi-City 371-0816, Japan
zhong@maebashi-it.ac.jp
[3] Dept. of Computer Science, University of Regina
Regina S4S 0A2, Canada
yyao@cs.uregina.ca

**Abstract.** Figural effects demonstrate that the influence on reasoning performance derives from the figure of the presented syllogistic arguments (Johnson-Laird and Bara, 1984). It has been reported that figure P-M/M-S is easier to reason with than figure M-P/S-M with syllogistic generation paradigm (Johnson-Laird, 1984), where M is the middle term, S is the subject and P is the predicate of conclusion, respectively. However, the figural effects are still unclear in syllogistic evaluation paradigm. In order to study such effects, we employed the figure M-P/S-M/S-P and the figure P-M/M-S/S-P syllogistic evaluation tasks with 30 subjects using eye-movement. The results showed that figural effects that the figure P-M/M-S/S-P was more cognitively demanding than the figure M-P/S-M/S-P, occurred in major premise and conclusion for the early processes, and in both premises and conclusion for late processes, rather than in minor premise reported by Espino *et al* (2005) that the figure P-M/M-S has less cognitive load than the figure M-P/S-M with generation paradigm. Additionally, pre-/post-conclusion viewing analysis found that for the inspection times of both premises the figure P-M/M-S/S-P took up more cognitive resources than the figure M-P/S-M/S-P when after viewing the conclusion. The findings suggested there were differences in figural effects between evaluation and generation paradigm.

## 1  Introduction

Syllogistic reasoning is a form of deductive reasoning in which a logical conclusion is drawn from premises (the major premise and the minor premise). For example,

Major premise: All humans are mortal.
Minor premise: Some animals are human.
Conclusion: Some animals are mortal.

Each of the three distinct terms, "human," "mortal," and "animal," represents a category. "Mortal" is the major term; "animal" is the minor term; the common

N. Zhong et al. (Eds.): BI 2009, LNAI 5819, pp. 106–114, 2009.

term "human" is the middle term(M) that links together the promises. "animal" is the subject(S) of the conclusion, and "mortal" is the predicate(P) of the conclusion. The arrangement of the middle term in each premise yields four types of scheme for reasoning, known as the figures of syllogism, that is, the figures M-P/S-M, P-M/S-M, M-P/M-S and P-M/M-S.

There are two research paradigms in studying how humans perform syllogistic reasoning: one is conclusion-evaluation paradigm, which requires subjects to decide whether a presented conclusion is deduced from the premises and the other is conclusion-generation paradigm, which requires subjects to generate the conclusion following the premises. Researchers have found that the evaluation task is more prone to conclusion-drive processing and the generation task may be viewed as a premise-driven processing [14].

Frase [9] performed earlier research about figures of syllogism with the conclusion-evaluation paradigm. The results show that the figure M-P/S-M/S-P takes significantly less time and produces fewer errors than the figure P-M/M-S/S-P. Mediated association theory [9] is proposed to explain the results. As showed in Fig. 1, the figure M-P/S-M/S-P is a forward-chaining paradigm with the response chaining of S-M-P, while the figure P-M/M-S/S-P is a backward-chaining paradigm with response chaining of P-M-S when inferring the conclusion with the direction of S-P. Late studies on syllogism have demonstrated that the figure of premises has strong influence on syllogistic reasoning behaviors [2,3,12].

Johnson-Laird and Bara [12] reported a study on syllogism with the conclusion-generation paradigm. They stated that figural effects occur when subjects integrate premises. The mental model theory (MMT) [13] suggests that the figure P-M/M-S where the middle terms are contiguous is less cognitively demanding than the figure M-P/S-M where the middle terms are not. Furthermore, the figure M-P/S-M has a response bias of the conclusion with direction of S-P, while the figure P-M/M-S has a response bias of P-S. Unlike MMT, the dual mechanism theory [5,8,10,19], proposes that there are two distinct cognitive systems underlying reasoning. System 1, which is shared with other animals, comprises a set of autonomous subsystems that include both innate input modules and domain-specific knowledge acquired by a domain-general learning mechanism, and is considered as associative processing. System 2, which is specific to humans and constrained by working memory capacity, is considered as reflective processing and participates

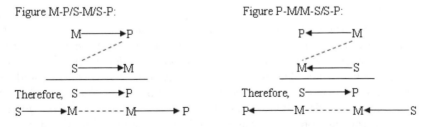

**Fig. 1.** Response chain of figure M-P/S-M/S-P and figure P-M/M-S/S-P

in human higher cognitive function. A neural imaging study explored dual mechanism theory in syllogistic reasoning [10]. It is found that a frontal-temporal system processes concrete materials while a parietal system processes abstract materials.

Espino et al [4] examined both the early and late processes in syllogistic reasoning in the conclusion-generation paradigm using eye-movement. The results showed that the effects of the difficulty of problems and the type of figure on syllogism are different. The the figure, but not the difficulty of syllogism, has effects on early processing; both the figure and the difficulty have effects on the late processing. The results also showed that figural effects mainly occur on minor premise; furthermore, the figure P-M/M-S is less time consuming in both early processing and late processing. Those results support MMT that the figure M-P/S-M creates an additional working memory load as the noncontiguous middle terms.

Stupple and Ball et al. [20] studied figural effects in syllogistic conclusion-evaluation paradigm. In their experiments, subjects can view only one of the three masked regions simultaneously and the results have yielded reliable support of MMT [13] that there are longer inspection time for the nonpreferred conclusion (the conclusion direction of P-S for the figure M-P/S-M and S-P for the figure P-M/M-S) than preferred one (the conclusion direction of S-P for the figure M-P/S-M and P-S for the figure P-M/M-S).

Previous studies reveal that figural effects are more explicit in conclusion-generation paradigm [12]. Those studies focus on belief bias effects in the evaluation paradigm [7,16]. Figural effects in the conclusion-evaluation paradigm are still not entirely clear [16]. In addition, previous studies typically used sequential premise presentation, which may burden the working memory load and constrain premises integration. In contrast, parallel premise presentation may be more effective for testing figural effects.

Eye-tracker has been widely used in cognitive studies and especially in studies of reading. Some authors argue that it can indicate the processes of mind [17,18]. The present study employed the conclusion-evaluation paradigm to explore the figural effects of the figure M-S/P-M/S-P and the figure P-M/M-S/S-P with parallel stimulus presentation method based on eye-movement.

## 2   Method

In the study, experiments were performed by using eye-tracker. The early processes, revealing the comprehension of an ROI, could be analyzed by the first fixation duration which denoted the duration of the given region of interest (ROI) from the time subject entered in it until he/she left it forwards or backwards for the first time. The late processes, revealing the reanalysis and integration of an ROI, could be analyzed by the sum of the duration of all the fixations except the first fixation duration, which was different from Espino's study where the late processes was defined as the sum of the duration of all the fixations made in a given part [4]. Pre-conclusion and post-conclusion viewing analysis [1] was used to test the impact of conclusion on figural effects in syllogistic evaluation paradigm. Data were analyzed using SPSS 15.0 software (http://www.spss.com/).

## 2.1   Subjects

Thirty paid undergraduate or graduate students from the Beijing University of Technology participated in the experiments (15 males and 15 females; aged $24.2 \pm 2.1$ years). All subjects were native Chinese speakers and right-handed, with normal or corrected-to-normal vision. None of the subjects reported any history of neurological or psychiatric diseases. All subjects gave informed consent.

## 2.2   Design and Stimuli

The experiments employed one-factor within subject design. Two conditions were the figure M-P/S-M/S-P and the figure P-M/M-S/S-P, respectively. Each subject received 90 figure M-P/S-M/S-P tasks and 90 figure P-M/M-S/S-P tasks. Examples were showed in Table 1. All the stimuli were selected or adapted from previous studies. The ROIs were major premise, minor premise, and conclusion, respectively.

**Table 1.** Examples of tasks (translated from Chinese)

| Figure M-P/S-M/S-P |
| --- |
| Major premise: All Greeks are the whites. |
| Minor premise: Some people are Greeks. |
| Conclusion: Some people are the whites. |
| Figure P M/M-S/S-P |
| Major premise: Some integers are natural numbers. |
| Minor premise: All natural numbers are real numbers. |
| Conclusion: Some real numbers are integers. |

## 2.3   Apparatus and Procedure

The syllogistic reasoning tasks were presented by Tobii Studio with the sampling rate of 60 Hz, and the monitor's resolution was $1024 \times 768$ pixel. Subjects were tested individually in a soundproof and comfortable room with soft light. All the sentences were presented in "Song" font with average length of 9 characters. The premises and conclusion were presented simultaneously and the procedure was showed in Fig. 2. Subject was seated in front of the monitor 60 cm away and the calibration of the eye tracker was performed. The syllogisms were presented one by one in a same pseudorandom order to each subject. Although subjects were given enough time to complete the tasks, they were instructed to judge whether the given conclusion was "true" and "false" instead of the logical terms "valid" and "invalid" by pressing a button as fast as possible.

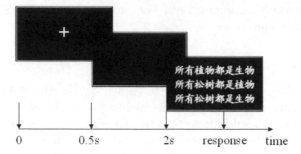

**Fig. 2.** Stimuli presentation (translation in English: All plants are creature; All pine trees are plants; All pine trees are creature)

## 3   Results

### 3.1   Results of Correctness and Response Time

Results of correctness and response time were showed in Fig. 3. The correctness was significantly higher for the figure M-P/S-M/S-P (0.80 $\pm$ 0.04) than that of the figure P-M/M-S/S-P (0.71 $\pm$ 0.12) ($F(1, 29) = 15.29, p < 0.001$). The results were consistent with the ones reported by Frase [9], but inconsistent with conclusion-generation paradigm [12,13]. The response time was significantly shorter for the figure M-P/S-M/S-P (8.82 $\pm$ 3.22) than that of the figure P-M/M-S/S-P (12.42 $\pm$ 5.55) ($F(1, 29) = 38.45, p < 0.001$).

### 3.2   Results of Early and Late Processes

The results of early and late processes analysis were showed in Table 2. The early processes were significantly longer for the major premise and conclusion of the figure P-M/M-S/S-P than that of the figure M-P/S-M/S-P ($F(1, 29) = 14.51, p < 0.005, F(1, 29) = 37.891, p < 0.001$, respectively). The difference of

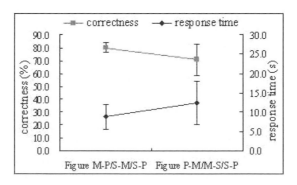

**Fig. 3.** The correctness and response time

**Table 2.** Early and late processes of premises and conclusion (unit: s)

| Tasks type | Early processes | Late processes |
|---|---|---|
| Figure M-P/S-M/S-P | | |
| major premise | $0.25 \pm 0.05$ | $1.87 \pm 1.01$ |
| minor premise | $0.26 \pm 0.06$ | $1.85 \pm 0.88$ |
| conclusion | $0.26 \pm 0.05$ | $0.89 \pm 0.44$ |
| Figure P-M/M-S/S-P | | |
| major premise | $0.27 \pm 0.05$ | $2.51 \pm 1.58$ |
| minor premise | $0.27 \pm 0.05$ | $2.43 \pm 1.24$ |
| conclusion | $0.29 \pm 0.05$ | $1.41 \pm 0.73$ |

the first fixation duration for the minor premise did not reach a statistical significance. The late processes were significantly longer for the major premise, the minor premise and the conclusion of figure P-M/M-S/S-P than that of figure M-P/S-M/S-P ($F(1, 29) = 19.331, p < 0.001; F(1, 29) = 25.281, p < 0.001; F(1, 29) = 48.128, p < 0.001$ respectively). The results might imply that both early and late processes happened in the major premise (premise 1), which was different from ones reported by Espino [4] that the figural effects appeared in the minor premise (premise 2).

### 3.3  Results of Pre-/Post-conclusion Viewing Analysis

We postulated that it was conclusion-driven processing in syllogistic evaluation paradigm. To clarify the effects of the conclusion on syllogistic reasoning, as showed in Table 3, we conducted an analysis of pre-conclusion and post-conclusion viewing analysis of the premises fixation durations [1]. The difference of pre-conclusion for the figure M-P/S-M/S-P and the figure P-M/M-S/S-P did not reach a statistical significance. The post-conclusion for the figure M-P/S-M/S-P was significantly less than the figure P-M/M-S/S-P ($F(1, 29) = 26.76, p < 0.001$). The results showed that the conclusion-driven processing might be applied in conclusion-evaluation, and figure P-M/M-S/S-P was more cognitively demanding than figure M-P/S-M/S-P when after viewing the conclusion.

**Table 3.** Fixation durations of pre-/post-conclusion viewing (unit: s)

| Inspection time | Figure M-P/S-M/S-P | Figure P-M/M-S/S-P |
|---|---|---|
| pre-con | $2.93 \pm 1.21$ | $2.99 \pm 1.22$ |
| post-con | $2.35 \pm 1.14$ | $3.48 \pm 1.81$ |

Notes: pre-con denotes the inspecting time of premises before the first fixation on the conclusion; post-con denotes the inspecting time of the premises after the first fixation on the conclusion.

# 4    Discussions

The present study addressed the figural effects in syllogistic reasoning with the conclusion-evaluation paradigm and simultaneous presentation of premises and conclusion using eye-movement technique. The eye-movement methodology could ensure reliable inspection time analysis. The statistical results of correctness and response times showed that the figure P-M/M-S/S-P was more cognitively demanding than the figure M-P/S-M/S-P. The result was consistent with ones reported by Frase [9]. At the same time, this result did not confirm the MMT prediction that the figure P-M/M-S is easier than figure M-P/S-M [12,13]. Our recent fMRI study about figural effects showed similar results that the figure P-M/M-S/S-P, which seemed backward-chaining processing, was more cognitively demanding than the figure M-P/S-M/S-P, which was a kind of forward-chaining processing [11]. The details will be discussed in the following analysis.

## 4.1    Early and Late Processes Analysis

The early processes implicated the initial presentation of problems and the late processes referred to the information integration. Both early and late processes showed that the figure P-M/M-S/S-P was more cognitively demanding than the figure M-P/S-M/S-P. The results did not support the critical MMT prediction that additional mental manipulations are required to align such noncontiguous middle terms for figure M-P/S-M [12]. Meanwhile, the result gave a hint that the processing might be different between the conclusion-evaluation and conclusion-generation paradigms [4].

It might be reasonable to consider following issues: (1) Conclusion-generation was considered to be premise-driven processing in which the contiguous middle terms of the figure P-M/M-S was less cognitively demanding. (2) Conclusion-evaluation was considered as conclusion-driven processing in which the conclusion might offer a heuristic strategy. Thus, the figure P-M/M-S/S-P was more cognitive load, like backward-chaining processing with the response chain of P-M-S. On the contrary, the figure M-P/S-M/S-P might be forward-chaining processing with response chain of S-M-P. (3)In Espino's [4] study the premises were displayed in a line, that is, the terms of S, M, P were displayed in a one-dimension view in which it might be more cognitively demanding and visual load when integrating the figure M-P/S-M with the noncontiguous middle terms than the figure P-M/M-S with contiguous middle terms; whereas, in our experiments the premises and the conclusion were displayed in three lines. The terms of S, M, P were displayed in a two-dimension view in which the differences in visual load were diminished. According to the analysis, the results were consistent with ones reported by Frase [9] that figure P-M/M-S/S-P was more cognitively demanding, and challenged the prediction of MMT.

## 4.2    Conclusion-Driven Processing Analysis

In order to clarify the impact of conclusion on figural effects in syllogistic reasoning, we did the pre-conclusion and post-conclusion viewing analysis. The results showed that before viewing the conclusion there was no difference between the figure M-P/S-M/S-P and the figure P-M/M-S/S-P in fixation duration; whereas after viewing the conclusion, the fixation duration of figure P-M/M-S/S-P was significantly longer than figure M-P/S-M/S-P. The results offered evidence that the processing in syllogistic evaluation paradigm might be a conclusion-driven pattern and figure P-M/M-S/S-P was more cognitively demanding than figure M-P/S-M/S-P when after viewing the conclusion.

According to Frase [9], the figure M-P/S-M/S-P might be a forward-chaining processing with the response chain of S-M-P; whereas the figure P-M/M-S/S-P might be a backward-chaining processing with the response chain of P-M-S. As a result, when subjects were required to evaluate the validity of figure P-M/M-S/S-P with the given conclusion S-P, it was relative difficult (longer response time and more errors) as it was easy to conclude the relationship of P-S just as formulated by MMT [12,13]. The results could be explained by the dual mechanism theory [5,8,10,19]. Dual mechanism theories have various flavors. The present study was consistent with the dual mechanism account developed by Newell and Simon [15]for the domain of problem solving that the heuristic and formal processes correspond to system 1 and system 2 respectively. Figure M-P/S-M/S-P automatically utilized situation-specific heuristics, which were based on background knowledge and experience. However, when conflict, between preferred conclusion direction of P-S and given conclusion direction of S-P for figure P-M/M-S/S-P, was detected, the formal (analytic) system was more active to control reasoning performance and to overcome the prepotent system 1. However, with the limit of working memory capacity there were longer response times and more errors for the figure P-M/M-S/S-P.

## 5    Conclusions

The present study addressed the reliable figural effects in syllogistic evaluation paradigm. From our experiments, one might conclude that there was conclusion-driven processing in evaluation paradigm, at the same time, the figural effects represented figure P-M/M-S/S-P was more cognitively demanding than figure M-P/S-M/S-P different from generation paradigm with premise-driven processing. The results might imply the mental model theory (MMT) was not suitable for syllogistic evaluation paradigm, whereas dual mechanism theory might shed a light on the situation, which still needed further experiments to confirm it.

## Acknowledgments

This work was supported in part by the National Natural Science Foundation of China (No. 60775039), a grant-in-aid for scientific research (No. 18300053) from the Japanese Ministry of Education, Culture, Sport, Science and Technology, and the Open Foundation of Key Laboratory of Multimedia and Intelligent Software Technology (Beijing University of Technology). The authors would like to thank Zhijiang Wang for his assistance in this study.

# References

1. Ball, L.J., Phillips, P., Wade, C.N., Quayle, J.D.: Effects of Belief and Logic on Syllogistic Reasoning. Experimental Psychology 53(1), 77–86 (2006)
2. Bara, B.G., Bucciarelli, M., Johnson-Laird, P.N.: Development of Syllogistic Reasoning. American Journal of Psychology 108, 157–193 (1995)
3. Bara, B.G., Bucciarelli, M., Lombardo, V.: Mental Model Theory of Deduction: A Unified Computational Approach. Cognitive Science 25, 839–901 (2001)
4. Espino, O., Santamaria, C., Meseguer, E., Carreias, M.: Early and Late Processes in Syllogistic Reasoning: Evidence for Eye-movements. Cognition 98, B1–B9 (2005)
5. Evans, J.S.B.T., Barston, J., Pollard, P.: On the Conflict between Logic and Belief in Syllogistic Reasoning. Memory Cognition 11, 295–306 (1983)
6. Evans, J.St.B.T., Handley, S.J., Harper, C.N.J., Johnson-Laird, P.N.: Reasoning about Necessity and Possibility: A Test of the Mental Model Theory of Deduction. Journal of Experimental Psychology: Learning, Memory, and Cognition 25, 1495–1513 (1999)
7. Evans, J.St.B.T., Handley, S.J., Harper, C.: Necessity, Possibility and Belief: A Study of Syllogistic Reasoning. Quarterly Journal of Experimental Psychology 54A, 935–958 (2001)
8. Evans, J.St.B.T.: Two Minds: Dual-Process Accounts of Reasoning. Trends in Cognitive Sciences 7, 454–459 (2003)
9. Frase, L.T.: Association Factors in Syllogistic Reasoning. Journal of Experimental Psychology 76, 407–412 (1968)
10. Goel, V., Buchel, C., Frith, C., Dolan, R.J.: Dissociation of Mechanisms Underlying Syllogistic Reasoning. NeuroImage 12, 504–512 (2000)
11. Jia, X.Q., Lu, S.F., Zhong, N., Yao, Y.Y., Li, K.C., Yang, Y.H.: Common and Distinct Neural Substrates of Forward-chaining and Backward-chaining Syllogistic Reasoning. In: International Conference on Complex Medical Engineering, ICME 2009 (2009)
12. Johnson-Laird, P.N., Bara, B.G.: Syllogistic Inference. Cognition 16, 1–61 (1984)
13. Johnson-Laird, P.N., Byrne, R.M.J.: Deduction. Erlbaum, Hove (1991)
14. Morley, N.J., Evans, J.St.B.T., Handley, S.J.: Belief Bias and Figural Bias in Syllogistic Reasoning. Quarterly Journal of Experimental Psychology 57A, 666–692 (2004)
15. Newell, A., Simon, H.A.: Human Problem Solving. Prentice- Hall, Englewood Cliffs (1972)
16. Quayle, J.D., Ball, L.J.: Working Memory, Metacognitive Uncertainty, and Belief Bias in Syllogistic Reasoning. Quarterly Journal of Experimental Psychology 53A, 1202–1223 (2000)
17. Rayner, K., Sereno, S.C., Morris, R.K., Schmauder, A.R., Clifton, C.: Eye movements and On-line Language Comprehension Processes. Language and Cognitive Processes 4, 21–50 (1989)
18. Rayner, K.: Eye movements in Reading and Information Processing: 20 Years of Research. Psychological Bulletin 124, 372–422 (1999)
19. Sloman, S.A.: The Empirical Case for Two Systems of Reasoning. Psychol. Bull 119(1), 3–22 (1996)
20. Stupple, E.J.N., Ball, L.J.: Figural Effects in a Syllogistic Evaluation Paradigm. Experimental Psy-chology 54(2), 120–127 (2007)

# Structured Prior Knowledge and Granular Structures

Qinrong Feng[1,2] and Duoqian Miao[2]

[1] School of Mathematics and Computer Science, Shanxi Normal University, Linfen, Shanxi, 041004, P.R.China
[2] Department of Computer Science and Technology, Tongji University, Shanghai, 201804, P.R.China

**Abstract.** In this paper, a hierarchical organization of prior knowledge based on multidimensional data model is firstly proposed, it is the basis of structured thinking. Secondly, a representation of granular structures based on multidimensional data model is also proposed, it can represents information from multiview and multilevel. Finally, the relation between structured prior knowledge and granular structures is analyzed.

## 1  Introduction

Granular computing is a general computation theory for effectively using granules such as classes, clusters, subsets, groups and intervals to build an efficient computational model for complex applications with huge amounts of data, information and knowledge [1]. It is also a way of thinking [2] that relies on the human ability to perceive the real world under various levels of granularity (i.e., abstraction).

Hobbs [3] stated that 'We perceive and represent the world under various grain sizes, and abstract only those things that serve our present interests. The ability to conceptualize the world at different granularities and to switch among these granularities is fundamental to our intelligence and flexibility. This enables us to map the complexities of real world into computationally tractable simpler theories'. Gordon et al. [4] pointed out that human perception benefits from the ability to focus attention at various levels of detail and to shift focus from one level to another. The grain size at which people choose to focus affects not only what they can discern but what becomes indistinguishable, thus permitting the mind to ignore confusing detail. Bradley C. Love [5] noticed that humans frequently utilize and acquire category knowledge at multiple levels of abstraction. Yao [2] proposed the basic ideas of granular computing, i.e., problem solving with different granularities. Granular computing, as a way of thinking, can captures and reflects our ability to perceive the world at different granularity and to change granularities in problem solving.

Although these scholars have noticed that human can solve problem at different levels of granularity, the reason of which has seldom been analyzed until now. If we know how the human intellect works, we could simulate it by machine.

N. Zhong et al. (Eds.): BI 2009, LNAI 5819, pp. 115–126, 2009.

In our opinion, this is concerned with human cognition, which leads to significant differences between human and machine in problem solving. Firstly, human can use relevant prior knowledge subconsciously in problem solving, but machine can't. For example, we notice that everyone can solve a given problem easily from his/her familiar fields at different levels of granularities. But (s)he even cannot solve a problem from his/her unfamiliar field at single level, not to mention at multiple levels. Secondly, human can use their relevant prior knowledge to generate a good 'structure' or problem representation in problem solving. As we all know that the famous story of young Gauss who gave the answer to the sum of all the numbers from 1 to 100 very quickly, not by very fast mental arithmetic but by noticing a pattern in the number sequence. Namely, that the numbers form pairs (1+100=101, 2+99=101, $\cdots$, 50+51=101). This example indicates that a good structuring, or representation of the problem helps considerably. In essence, the so called 'good' structure of a problem is a hierarchical structure induced from it. Thirdly, J.Hawkins and S.Blakeslee [6] pointed out that there have fundamentally different mechanisms between human brain and machine. The mechanism of human brain is that it retrieves the answers stored in memory a long time ago, but not "compute" the answers to a problem as machine. This indicates that human usually search a relevant or similar answer to the solved problem from his memory in problem solving, or we can say that it depends on relevant prior knowledge to solve problem for human. Maybe these can be used to interpret why human can focus on different levels during the process of problem solving, but machine cannot.

In this paper, the importance of hierarchical structured prior knowledge in granular computing is stressed firstly, where prior knowledge is extended to a broader sense, which includes domain knowledge. A nested and hierarchical organization of prior knowledge based on multidimensional data model is proposed, it is the basis of structured thinking [7]. Secondly, multidimensional data model was introduced into granular computing, it can represent information from multiview and multilevel, and can be used as a representation model of granular structures. Finally, the relation between structured prior knowledge and granular structures is analyzed.

## 2    Prior Knowledge

In this section, we will give a brief introduction to prior knowledge, and propose an organization of prior knowledge based on multidimensional data model. We also point out that structured prior knowledge is the basis of structured thinking. The main role of structured prior knowledge is to provides humans with a much greater control over the solved problem.

### 2.1    An Introduction to Prior Knowledge

Prior knowledge is all the knowledge you've acquired in your lifetime. It includes knowledge gained from formal and informal instruction. Prior knowledge has the same meaning as background knowledge, previous knowledge, personal knowledge, etc. In this paper, we will extend the concept "prior knowledge" to a broader sense, which also includes domain knowledge.

Prior knowledge can help us all of the time. When we do something for the first time, we will feel hard because we don't have much prior knowledge about it. After we do something several times and accumulate a lot of relevant prior knowledge, we will feel easier. This indicates that prior knowledge is very helpful to human problem solving.

In practice, prior knowledge is valuable to be incorporated into a practical problem solving. Yu et al. [8] noticed the necessity of extra information to problem solving, and provided a mathematical expression

$$Generalization = Data + knowledge \qquad (1)$$

This formula indicates that if you want to solve a problem at higher levels or multiple levels of abstraction, you will have to add extra knowledge to the solved problem. In this formula, data can be obtained from the solved problem. But what's the knowledge here? In our opinion, the knowledge here is what we call prior knowledge in this paper.

Prior knowledge can be obtained by learning. When we learn new knowledge, we will assimilate it and make sense of it by connecting it to what we have already known, or we can say that we will incorporate it into our existing knowledge structure subconsciously. As stated in [9]: "Our representation of the world is not necessarily identical to the actual world. We modify information that is received through our senses, sharpening, selecting, discarding, abstracting, etc. So our internal representation of the world is really our own construction. In other words, human brain is not a sponge that passively absorbs information leaking out from the environment. Instead, they continually search and synthesize." Which structure does prior knowledge be organized in human brain? And which organization of prior knowledge is suitable for granular computing particularly? These will be discussed in the next subsection.

## 2.2 Structure of Prior Knowledge

Some researchers have noticed the importance of knowledge structure. Mandler [10] pointed out that meaning does not exist until some structure, or organization, is achieved. In [11], the authors pointed out that knowledge structure is a structured collection of concepts and their interrelationships, it includes two dimensions: multilevel structure and multiview structure. This is in accordance with granular structures. In [6], the authors pointed out that "the cortex's hierarchical structure stores a model of the hierarchical structure of the real world. The real world's nested structure is mirrored by the nested structure of your cortex". In an exactly analogous way, our memories of things and the way our brain represents them are stored in the hierarchical of the cortex. In [12], the authors mentioned that humans and other species represent knowledge of routine events or stereotypical action sequences hierarchically. These indicate that prior knowledge can be organized as a hierarchical structure. But which hierarchical structure can reflect prior knowledge more intuitively?

Quillian [13,14,15] pointed out that prior knowledge is stored in the form of semantic networks in human brain. A semantic network or knowledge structure is created with three primitives: concepts, relations, and instances. Yao [16] pointed out human thought and knowledge is normally organized as hierarchical structures, where concepts are ordered by their different levels of specificity or granularity. A plausible reason for such organizations is that they reflect truthfully the hierarchical and nested structures abundant in natural and artificial systems. Human perception and understanding of the real world depends, to a large extent, on such nested and hierarchical structures.

Of course, we don't know the genuine organization of prior knowledge in human brain until now. But from the above we know that prior knowledge should be organized as a nested and hierarchical structures, which is helpful to human problem solving.

### 2.3   Organization of Prior Knowledge

Maybe there have a lot of knowledge stored in human brain, but only a little part of which is relevant to the solved problem in problem solving. So we should only extract the relevant part of prior knowledge and reorganize them as a nested and hierarchical structure in problem solving. Hierarchical structures not only make a complex problem more easily understandable, but also lead to efficient solutions.

In this subsection, we will provide a nested and hierarchical organization of prior knowledge relevant to the solved problem, which is suitable for problem solving particularly. We will reorganize relevant prior knowledge from one point of view with multiple levels of granularity as a concept hierarchy, and reorganize relevant prior knowledge from multilevel and multiview as a multidimensional data model.

The formal use of concept hierarchies as the most important background knowledge in data mining was introduced by Han, Cai and Cercone [18]. And a multidimensional data model can be regarded as a combination of multiple concept hierarchies. So it is rational to represents prior knowledge as concept hierarchies or multidimensional data model.

**Concept hierarchy.** Concepts are the basic unit of human thoughts and play a central role in our understanding of the world. Human usually has a rich clustering of concepts for knowledge from his familiar field, in which each concept is related to many other concepts, and the relationships between concepts are clearly understood. Concepts are arranged hierarchically using umbrella concepts to more tightly relate them. The concept hierarchy is such an example.

In what follows, an organization of prior knowledge from one point of view with multiple levels of granularity is proposed based on concept hierarchy. In order to satisfy the need of nested and hierarchical structure, we will organize relevant prior knowledge as a concept hierarchy in the form of tree in this paper.

A concept hierarchy is a graph whose nodes represent concepts and whose arcs represent partial order relation between these concepts. In a concept hierarchy, the meaning of a concept is built on a small number of simpler concepts, which in turn is defined at a lower level using other concepts.

Concept hierarchies are used to express knowledge in concise and high-level terms. As P.Witold [17] pointed out that granular computing is an information processing pyramid, concept hierarchy tree just has the form of pyramid, where more nodes at lower level and they are all specific concepts, less nodes at higher level and they are abstract concepts.

In [1], the authors pointed out that the granulation process transforms the semantics of the granulated entities. At the lowest level in a concept hierarchy, basic concepts are feature values available from a data set. At a higher level, a more complex concept is synthesized from lower level concepts (layered learning for concept synthesis).

In [19], the authors stated "Concepts are not isolated into the human cognitive system. They are immersed into a hierarchical structure that facilitates, among others, classification tasks to the human cognitive system. This conceptual hierarchy expresses a binary relationship of inclusion defined by the following criterions:

*Inclusion criterion* : Each hierarchy node determines a domain included into domain of its father node. Each hierarchy node determines a domain that includes every domain of its son nodes.

*Generalisation-Specialisation criterion*: Every node in the hierarchy, has differentiating properties that make it different from its father node, if it exists, and from the others son nodes of its father, if they exist. "

Concept hierarchies may be defined by discretizing or grouping data, that is, discretizing numerical data into interval, and grouping categorical data into a generalized abstract concept. A total or partial order can be defined among groups of data.

For example, we can evaluate the ability of a person from multiview, such as 'education', 'vocation', 'income', etc. And, for each view, e.g., 'education', we can regard it from multilevel by relying on our relevant prior knowledge. For example, prior knowledge about 'education' can be organized as the following concept hierarchy.

**Fig. 1.** Concept hierarchy of 'Education'

The concept hierarchy illustrated as figure 1 possesses a nested and hierarchical structure, where every node represent a concept, and arcs represent partial order relation between these concepts. Nodes at lower level are represent specific concepts, and those at higher level are represent abstract concepts.

Different people may own different prior knowledge, and different people may also have different preferences. Thus, for a given problem, different people will construct different concept hierarchy. To illustrate this, we still take 'education' for example. If you are a manager of some college and university, or a manager of a science institute, your prior knowledge relevant to 'education' may be structured as figure 1. But in common people's opinion, 'undergraduate' should also belong to high education. Thus, prior knowledge is relevant to the context of a problem.

A concept hierarchy can only organizes relevant prior knowledge from one particular angle or point of view with multiple levels of granularity, but it can not organizes those from multiview.

In the next section, we will borrow the concept of multidimensional data model from data warehousing, and provide an organization of prior knowledge. This organization can represent relevant prior knowledge from multilevel and multiview.

**Multidimensional data model.** In 1969, Collins and Quillian [15] made a typical experiment to prove prior knowledge that stored in long-term memory are in network architecture. This experiment suggests that people organize knowledge structurally and stored the features of the concept in different levels of the hierarchical architecture.

How can we represent this network architecture intuitively? In this subsection, we will organize multiple concept hierarchies as an organic whole, which is represented by a multidimensional data model. This organization can not only make relevant prior knowledge more easily understandable, but also represents them from multiview and multilevel intuitively.

Multidimensional data model [20,21] is a variation of the relational model that uses multidimensional structures to organize data and express the relationships among data, in which the data is presented as a data cube, which is a lattice of cuboid. A multidimensional data model includes a number of dimensions that each includes multiple levels of abstraction defined by concept hierarchy. Thus, a multidimensional data model can be treated as a combination of multiple concept hierarchies, and it can represent data from multiview and multilevel. This organization provides users with the flexibility to view data from different perspective. Based on the hierarchical structure of multidimensional data model, it is possible to "scan" a data table from different levels of abstraction and different dimensions.

In a multidimensional data model, each dimension can be represented by a concept hierarchy, which can represents a problem from a particular angle. Concept hierarchies of multiple dimensions be organized as a multidimensional data model, which can represents data from multiview and multilevel. In fact, multidimensional data model itself is a well-organized network structure, it can reflect relevant prior knowledge intuitively and completely.

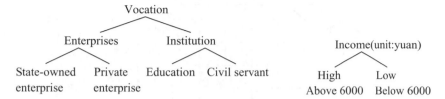

**Fig. 2.** Concept hierarchy of 'Vocation'    **Fig. 3.** Concept hierarchy of 'Income'

Next we will illustrate the ability of multidimensional data model represents prior knowledge from multiview and multilevel through an example. As the above mentioned, we can evaluate the ability of a person from multiview, such as 'education', 'vocation', 'income', etc. And for each view, we can regard it from multilevel by relying on our relevant prior knowledge.

We assume that prior knowledge about 'vocation' and 'income' be structured as concept hierarchies illustrated as figure 2 and figure 3, respectively. Then we will organize prior knowledge about 'education', 'vocation', and 'income' as a multidimensional data model, which can used to evaluate a person from multiview and multilevel.

For simplicity, we will denote those concepts in figure 1, figure 2 and figure 3 by some symbols. Such as denote 'education' by 'A', 'vocation' by 'B' and 'income' by 'C'. The correspondence between concepts and symbols is as follows.

'Education' ↔ 'A'
'High education' ↔ 'A1'        'Low education' ↔ 'A2'
'Doctoral student' ↔ 'A11'     'Postgraduate student' ↔ 'A12'
'Undergraduate' ↔ 'A21'        'Others' ↔ 'A22'
where 'A11', 'A12', 'A21', 'A22' are specific concepts, and 'A1' is abstracted from 'A11' and 'A12', 'A2' is abstracted from 'A21' and 'A22'.

'Vocation' ↔ 'B'
'Enterprises' ↔ 'B1'                    'Institution' ↔ 'B2'
'State-owned enterprise' ↔ 'B11'        'Private enterprise' ↔ 'B12'
'Education' ↔ 'B21'                      'Civil servant' ↔ 'B22'
Similarly, 'B11', 'B12', 'B21', 'B22' are specific concepts, and 'B1' is abstracted from 'B11' and 'B12', 'B2' is abstracted from 'B21' and 'B22'.

'Income' ↔ 'C'
'High income' ↔ 'C1'    'Low income' ↔ 'C2'

The multidimensional data model is organized from the above three concept hierarchies as figure 4, which can be used to evaluate a person from multiview (education, vocation and income) and multilevel. For example, the shadow cell in the data cube illustrated as figure 4 represents persons with high education, work as a civil servant and have a high income. Each concept hierarchy corresponds a dimension in figure 4.

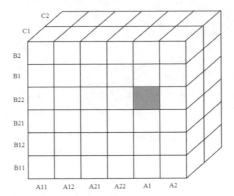

**Fig. 4.** Multidimensional data model

# 3   Granular Structure

In this section, we will give a brief introduction to granular structure, and present
a representation of granular structure from multiview and multilevel. This representation can be used to represent information from multi-granularity in granular
computing.

## 3.1   What Is Granular Structure

A central notion of granular computing is multilevel granular structures, which
consists of inter-connected and inter-acting granules, families of granules interpreted as levels of differing granularity, and partially ordered multiple levels
known as hierarchical structures. They are the results of a structured understanding, interpretation, representation, and description of a real-world problem
or system. Granular structures provide structured descriptions of a system or a
problem under consideration.

Granular computing emphasizes on structures. The study of granular computing depends crucially on granular structures that represent reality through
multilevel and multiview. It seems that hierarchical granular structure is a good
candidate for developing a theory of granular computing.

## 3.2   Representation of Granular Structure

In granular computing, we usually represent a problem from one particular angle
or point-of-view with multiple levels of granularity by a hierarchy. However, the
conceptualization of a problem through multiple hierarchies (i.e., multiview) and
multilevel in each hierarchy is general and flexible. A complete understanding
of a problem requires a series of granular structures that should reflect multiple
views with multiple levels [16,22]. Thus, granular structures need to be modeled
as multiple hierarchies and multiple levels in each hierarchy. That is, granular
structures should reflect multiview and multilevel in each view. But there has
no effective model to represent granular structure in existing literature.

We notice that the structure of multidimensional data model is consistent with the needs of granular structure. A multidimensional data model includes multiple dimensions, and each dimension includes multiple levels of abstraction. So each dimension in a multidimensional data model corresponds to a view with multiple levels of granularity in granular structures. Thus, we will represent a problem from multiview and multilevel by a multidimensional data model, which use multidimensional structure to organize data and express the relationships among data. This representation will facilitate the process of problem solving. Moreover, concept hierarchy and multidimensional data model can not only to elicit the content of knowledge, but also its structure.

Granular structures provide descriptions of a system or a problem under consideration. Yao [23] pointed out that granular structures may be accurately described as a multilevel view given by a single hierarchy and a multiview understanding given by many hierarchies. But he didn't mention how to organize these multiple hierarchies as an organic whole.

In what follows, we will present multidimensional data model representation of granular structures through an example.

**Example 1.** Representing granular structures of the following problem by a multidimensional data model, where the problem is provided by table 1.

This problem is provided by a table, where every column represents a particular view of the problem. So every column can be represented by a concept hierarchy, as shown in figure 1, figure 2, and figure 3. So this problem can be represented as a multidimensional data model by organizing these concept hierarchies as an organic whole, as shown in figure 5. The objects in cells of data cube as in figure 5 are satisfy properties determined by corresponding coordinate. Or we can say that the objects in cells of data cube is the extension of a concept, and the intension of the concept is designated by coordinate of the corresponding cell.

We can obtain the multidimensional data model representation of table 1 as figure 6 by combining table 1 and granular structure illustrated as figure 5. Or in other words, we can obtain multidimensional data model as figure 6 by loading data in table 1 to granular structure as figure 5. For example, the object 4, 5, 10 in data cube possesses the properties 'A2', 'B1' and 'C2' simultaneously, that is, these objects possess properties as 'Low education', work in 'Enterprises' and 'Low income'.

For a given table, we can generalize its every attribute to a concept hierarchy tree, and organize these concept hierarchy trees as a multidimensional data model. In essence, the multidimensional structure of this multidimensional data model is the granular structure hid in the given table. Thus a given table can be generalized to multiple tables with different degrees of abstraction by combining this table and granular structures hid in it. These tables with different degree of abstraction is the basis of structured problem solving and structured information processing.

**Table 1.** Training dataset

| U | Education | Vocation | Income(unit:yuan) |
|---|---|---|---|
| 1 | Doctoral student | Private enterprise | High |
| 2 | Postgraduate student | State-owned enterprise | High |
| 3 | Others | Education | Low |
| 4 | Undergraduate | Private enterprise | Low |
| 5 | Undergraduate | State-owned enterprise | Low |
| 6 | Postgraduate student | State-owned enterprise | Low |
| 7 | Undergraduate | State-owned enterprise | High |
| 8 | Undergraduate | Civil servant | Low |
| 9 | Doctoral student | Education | Low |
| 10 | Others | State-owned enterprise | Low |

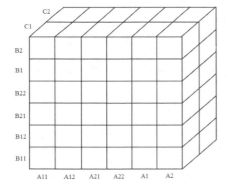

**Fig. 5.** Granular structure of table 1

**Fig. 6.** Multidimensional data model of table 1

## 4    Relation between Structured Prior Knowledge and Granular Structure

Every one may be possess much knowledge, which usually be organized as a complicated network architecture in human brain. But prior knowledge is always related to a problem. When we solve a problem, we only need to consider prior knowledge relevant to it. Prior knowledge relevant to the solved problem will be extracted from the complicated network structure, and be reorganized as a nested and hierarchical structure, we call it structured prior knowledge, which will helpful to human problem solving.

Granular structure is a central notion of granular computing, it can represents a problem from multilevel and multiview. Granular structure can make implicit knowledge explicit, make invisible knowledge visible, make domain-specific knowledge domain-independent and make subconscious effects conscious [7]. In essence, granular structure is a structure hid in the solved problem.

As stated above, we can see that prior knowledge and granular structure are related together tightly by the solved problem. There have many points of similarity between prior knowledge and granular structure. For instance, they have a similar structure, that is, they all can be modeled as a nested and hierarchical structure, and each hierarchical structure also includes multiple levels of abstraction. They also have a similar ability of representation, that is, they all can represent a problem from one view with multiple levels of abstraction or from multiview and multilevel. In fact, Granular structures is an intuitive reflection of structured prior knowledge relevant to the solved problem, and the construction of granular structures is need the guidance of relevant prior knowledge.

# 5   Conclusion

In this paper, a nested and hierarchical organization of prior knowledge based on multidimensional data model is proposed, it is the basis of structured thinking. A representation of granular structures based on multidimensional data model is also proposed, it can represents a problem from multiview and multilevel. Finally, the relation between structured prior knowledge and granular structure is analyzed.

From the discussion, we conclude that the reason of human intelligence can solve problem at different levels of granularities is that human can not only use hierarchically organized relevant prior knowledge subconsciously, but also use structured prior knowledge to reorganize the solved problem as a good representation (granular structure) in problem solving.

**Acknowledgments.** This work was supported by Youth Technology Research Foundation of Shanxi Province(2008021007), the Natural Science Foundation of Shanxi Normal University(YZ06001) and National Natural Science Foundation of China (Serial No. 60775036).

# References

1. Bargiela, A., pedrycz, W.: The roots of granular computing. In: Proceedings of the IEEE international conference on granular computing, pp. 806–809 (2006)
2. Yao, Y.Y.: A partition model of granular computing. In: Peters, J.F., Skowron, A., Grzymała-Busse, J.W., Kostek, B.z., Świniarski, R.W., Szczuka, M.S. (eds.) Transactions on Rough Sets I. LNCS, vol. 3100, pp. 232–253. Springer, Heidelberg (2004)
3. Hobbs, J.R.: Granularity. In: Proceedings of the Ninth International Joint Conference on Artificial Intelligence, pp. 432–435 (1985)
4. Mccalla, G., Greer, J., Barrie, B., Pospisil, P.: Granularity Hierarchies. In: Computers and Mathematics with Applications: Special Issue on Semantic Networks (1992)
5. Love, B.C.: Learning at different levels of abstraction. In: Proceedings of the Cognitive Science Society, USA, vol. 22, pp. 800–805 (2000)
6. Hawkins, J., Blakeslee, S.: On Intelligence. Henry Holt and Company, New York (2004)

7. Yao, Y.Y.: The art of granular computing. In: Kryszkiewicz, M., Peters, J.F., Rybiński, H., Skowron, A. (eds.) RSEISP 2007. LNCS (LNAI), vol. 4585, pp. 101–112. Springer, Heidelberg (2007)
8. Yu, T., Yan, T., et al.: Incorporating prior domain knowedge into inductive machine learning (2007), http://www.forecasters.org/pdfs/DomainKnowledge.pdf
9. A brief history of cognitive psychology and five recurrent themes, http://road.uww.edu/road/eamond/351-Cognitive_Psychology/Slides-Printed/L02-Brief_History_and_5_themes.pdf
10. Mandler, Jean, M.: Stories: The function of structure, Paper presented at the annual convention of the American Psychological Association, Anaheim, CA (1983)
11. Chow, P.K.O., Yeung, D.S.: A multidimensional knowledge structure. Expert Systems with Applications 9(2), 177–187 (1995)
12. Mesoudi, A., Whiten, A.: The hierarchical transformation of event knowledge in human cultural transmission. Journal of cognition and culture 4.1, 1–24 (2004)
13. Quillian, M.R.: Word concepts: A theory and simulation of some basic semantic capabilities. Behavioral Sciences 12, 410–430 (1967)
14. Quillian, M.R.: Semantic meaning. In: Minsky, M. (ed.) Semantic information processing. MIT Press, Cambridge (1968)
15. Quillian, M.R.: The teachable language comprehender. In: Communications of the Association for Computing Machinery, vol. 12, pp. 459–475 (1969)
16. Yao, Y.Y.: Granular computing for Web intelligence and Brain Informatics. In: Proceedings of the IEEE/WIC/ACM International Conference on Web Intelligence, pp. xxxi–xxiv (2007a)
17. Pedrycz, W.: Granular computing: An introduction. In: Joint 9th IFSA World Congress and 20th NAFIPS International Conference, vol. 3, pp. 1349–1354 (2001)
18. Han, J., Cai, Y., Cercone, N.: Data-driven discovery of quantitative rules in relational database. IEEE Transactions on Knowledge and Data Engineering 5(1), 29–40 (1993)
19. de Mingo, L.F., Arroyo, F., et al.: Hierarchical Knowledge Representation: Symbolic Conceptual Trees and Universal Approximation. International Journal of Intelligent Control and Systems 12(2), 142–149 (2007)
20. Singh, M., Singh, P., Suman: Conceptual multidimensional model. In: Proceedings of world academy of science, engineering and technology, vol. 26, pp. 709–714 (2007)
21. Pedersen, T.B., Jensen, C.S.: Multidimensional database technology. Computer 34(12), 40–46 (2001)
22. Yao, Y.Y.: A Unified Framework of Granular Computing. In: Pedrycz, W., Skowron, A., Kreinovich, V. (eds.) Handbook of Granular Computing, pp. 401–410. Wiley, Chichester (2008a)
23. Yao, Y.Y.: Human-inspired granular computing. In: Yao, J.T. (ed.) Novel Developments in Granular Computing: Applications for Advanced Human Reasoning and Soft Computation (2008b)

# Multisensory Interaction of Audiovisual Stimuli on the Central and Peripheral Spatial Locations: A Behavioral Study

Qi Li [1,3], Jingjing Yang[1], Noriyoshi Kakura[2], and Jinglong Wu[1,4]

[1] Graduate School of Natural Science and Technology, Okayama University,
Okayama, Japan
[2] Graduate School of Engineering, Kagawa University, Takamatsu, Japan
[3] School of Computer Science and Technology, Changchun University of Science
and Technology, Changchun, China
[4] The International WIC Institute, Beijing University of Technology, China
wu@mech.okayama-u.ac.jp

**Abstract.** Much knowledge has been gained about how the brain is activated on audiovisual interaction. However, few data was acquired about the relationship between audiovisual interaction and spatial locations of audiovisual stimuli. Here, we investigated the multisensory interaction of audiovisual stimuli presented on central and peripheral spatial locations. Firstly, we determined the maximal eccentricity of peripheral spatial location on where stimuli were presented which was 30°. Second, the results of audiovisual interaction showed that the interaction of visual and auditory can have dramatic effects on human perception and performance. Moreover, this effect depended on the spatial locations of presented audiovisual stimuli.

## 1 Introduction

Recognizing an object requires one to pool information from various sensory modalities, and to ignore information from competing objects [1]. It is well documented that combining sensory information can enhance perceptual clarity and reduce ambiguity about the sensory environment [2]. For example, multisensory information can speed reaction times [3, 4, 5], facilitate learning [6] and change the qualitative sensory experience [2]. Typically, these perceptual enhancements have been found to depend upon the stimuli being within close spatial and temporal proximity, although interactions have been observed in some cases even with significant spatial disparities between the visual and auditory stimuli [7]. In addition, visual information dominates auditory spatial information in the perception of object location [8, 9]. Perceptual temporal acuity in the visual system is relatively poor compared with the auditory system [7, 10].

Some studies have shown that the reaction times to visual stimuli slowed with the eccentricity increase of stimuli [13]. And the capability of auditory localization also became low with the eccentricity increase of stimuli [14]. However, whether the eccentricity of stimuli affected the multisensory interaction remain unclear.

N. Zhong et al. (Eds.): BI 2009, LNAI 5819, pp. 127–134, 2009.

In present study, we mainly investigated whether multisensory interaction of audiovisual stimuli presented on the central and peripheral spatial locations was consistent. In order to gain maximal difference of audiovisual interaction between on the central and peripheral spatial locations, we conducted a pre-experiment prior to all experiments and determined the eccentricity of peripheral spatial location on where stimuli were presented.

## 2 Method

### 2.1 Subjects

Thirteen volunteers (22 to 31 years, mean: 24.8 years, both males), students from the Kagawa University participated in the experiment. All had normal or corrected-to-normal vision and normal hearing and all were right handed. The experimental protocol was approved by the ethics committee of Kagawa University. After receiving a full explanation of the purpose and risks of the research, subjects provided written informed consent for all studies as per the protocol approved by the institutional research review board.

### 2.2 Stimuli

The experiment contained three stimulus types which are unimodal visual (V) stimuli, unimodal auditory (A) stimuli, and multimodal audiovisual (VA) stimuli. Each type had two subtypes of standard stimuli and target stimuli.

The unimodal V standard stimulus was a white, horizontal, square wave grating (subtending a visual angle of approximately 3°). The unimodal A standard stimulus was of a 1600 Hz tone with linear rise and fall times of 5ms and an intensity of 65 dB. The multimodal VA standard stimulus consisted of the simultaneous presentation of both unimodal A and V standard stimuli at a spatially congruent location. The duration of each type of stimulus was 105 ms.

The unimodal V and A target stimuli were very similar to the unimodal V and A standard stimuli, but contained a distance of no stimulation exhibition time half way through the standard stimuli which subjectively made the stimulus appear to flicker (V target stimulus) or stutter (A target stimulus). The difficulty of the target stimulus detection can be adjusted for each subject during a training session prior to the experiment by selecting the different distance of no stimulation exhibition time halfway so that the accuracy in detecting the target stimuli was 80%. The V target stimulus contained four different distances of no stimulation exhibition time halfway which were 25ms, 35ms, 45ms, and 55ms. The A target stimulus also contained four different distances of no stimulation exhibition time halfway which were 15ms, 25ms, 35ms and 45ms. The multimodal VA target stimulus consisted of the simultaneous presentation of both unimodal A and V target stimuli at a spatially congruent location. The visual detection capability was degressive with the increase of the eccentricity of presented location. We conducted a pre-experiment prior to all experiments, in order to determine the max eccentricity of peripheral spatial location on where stimuli were presented. The pre-experiment contained 320 V standard stimuli and 320 V target stimuli. The V standard and target stimuli were presented with equal probability at an eccentricity of ±7.5°, ±10°, ±12.5°, ±15°, ±30°, ±45°, ±60°, ±75° along the horizontal meridian to the both sides of the central fixation point. The task to subjects was to respond by pressing a button with the right hand as quickly as possible if the V target stimuli appeared, but to withhold a response to the V standard stimuli. As shown in Figure 1, if the eccentricity exceeded 30°, the accuracies started to drop. Because the accuracies from eccentricity 45° were lower than 75%, therefore we determined the eccentricity of 30° as the peripheral spatial location on where stimuli were presented.

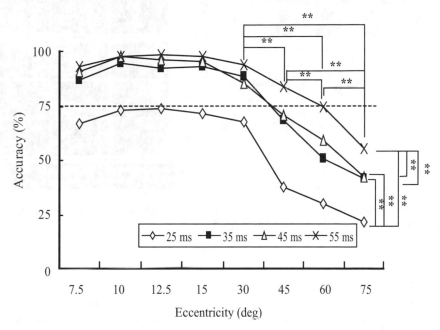

**Fig. 1.** The accuracies to V target stimuli with a 25ms, 35ms, 45ms or 55ms distance of no stimulation exhibition time halfway at the eccentricity of ±7.5°, ±10°, ±12.5°, ±15°, ±30°, ±45°, ±60°, ±75° along the horizontal meridian to the both sides of the central fixation point (** p < 0.001)

## 2.3 Task and Procedure

Experiments were conducted in a dark and sound-attenuated room. The stimulus presentation and data acquisition were controlled using the Presentation (Version 0.61) experimental software. The V stimuli were presented on a 17-in display which was placed directly in front of the subject's head at a distance of 30ms. The A stimuli were delivered through three speakers which were placed before the display at eccentricity of -30°, 0°, +30° (Figure 2A ).

The experiment had 15 sessions. Each session consisted of 60 unimodal A stimuli, 60 unimodal V stimuli, and 60 multimodal VA stimuli. All stimuli were randomly presented, and each type of stimulus was presented with equal probability on the left (-30°), right (+30°) and central (0°) locations. The target stimulus frequency was 20% in each stimulus type. Figure 2B showed the all stimulus types.

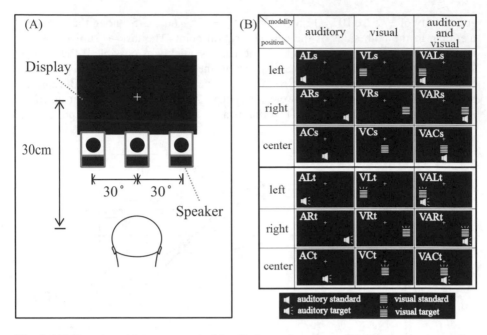

**Fig. 2.** (A) Experimental arrangement of the display and speakers. (B) All stimulus types. E.g. ALs: auditory standard stimulus presented on left side.

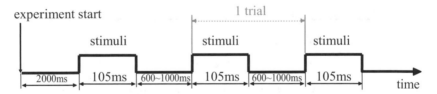

**Fig. 3.** Schematic of the trail sequence

As shown in Figure 3, after 2000ms of experiment onset, first stimulus was presented for 105ms on the left, right or central location. Following a random interstimulus interval of 600-1000ms, another stimulus was presented.

Throughout the experiment, the subjects were required to fixate on a centrally presented fixation point. The subject's task was to detect all target stimuli anywhere as quickly and accurately as possible and to respond by pressing the left key of a computer mouse. The subjects were allowed to take a 2-min break between sessions. At the beginning of experiment, the subjects were given a number of practice trials to ensure that they understood the paradigm and were familiar with the stimuli.

## 2.4 Data Analysis

Reaction times (RTs) and hit rates (HRs) for the target stimuli were computed separately for each stimulus type and location. The RTs and HRs were analyzed using a repeated-measures ANOVA with stimulus modality (V, A and VA) and stimulus location (left, right and central) as the subject factors.

## 3 Results

### 3.1 Response Times

Fig.4. showed the RTs of V, A and VA stimuli on left, right and central locations. The ANOVA analysis of variance revealed a significant effect of the main factors, stimulus modality [F (2, 48) =27.4, P<.001] and stimulus location [F (2, 48) =17.0, P<.001]. There was a significant interaction between stimulus modality and stimulus location [F (4, 48) =4.5, P<.05]. For the stimulus modality, RTs to VA target stimuli were significantly faster than that of unimodal V target stimuli and unimodal A target stimuli anywhere. For the stimulus location, RTs to V and VA target stimuli on the central location were faster than that on the left and right location.

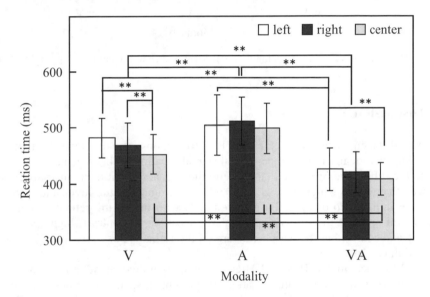

**Fig. 4.** Reaction times of V, A and VA stimuli on left, right and central locations (** p <0.001)

### 3.2 Hit Rates

Fig. 5 showed the HRs of V, A and VA stimuli on left, right and central locations. The ANOVA analysis of variance revealed that a significant effect of the main factors on stimulus modality [F (2, 48) =60.4, p<.001]. For the stimulus modality, HRs to A and VA target stimuli were more accurate than that of unimodal V target stimuli. For the stimulus location, HRs to V target stimuli on the central location was more accurate than that on the left and right location.

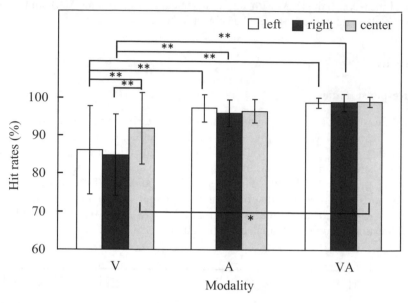

**Fig. 5.** Hit Rates of V, A and VA stimuli on left, right and central locations (* p < 0.05, ** p < 0.001)

## 4 Discussion

We found the enhancement of RTs and HRs to response the unimodal V stimuli on central spatial location than on peripheral spatial locations. This enhancement attribute to the difference of the cell structure in the retinae. Some neuroanatomy studies showed that the Pβ ganglion cells in the retinae dominate at the fovea and there is a decrease in the ratio of Pβ to Pα ganglion cells with increasing retinal eccentricity, the central parts of the retina is particularly sensitive to contrast, high spatial frequency and colour, but the peripheral parts of the retina is possesses a high temporal resolution [15,16].

No difference in the RTs and HRs response to unimodal auditory stimuli was found on between central spatial location and peripheral spatial locations. This results supported that spatial perception in the auditory system were relatively stolid [7, 10].

The responses to VA target stimuli were faster than those to unimodal V target stimuli and unimodal A target stimuli on each of three spatial locations (left, right, and central). This finding is consistent with some previous studies, and supported that the multisensory audiovisual information speeded reaction times [3, 4, 5]. Moreover, the RTs to responses the VA target stimuli presented on central spatial locations was shorter than that presented on peripheral spatial locations. This supports the hypothesis that the spatial location of presented stimuli affected audiovisual interaction.

In addition, we did not found HRs enhancement between unimodal A stimuli and multimodal VA stimuli. And the HRs to response the unimodal A stimuli was above 90%. We speculated that it is enough to subjects to detect the VA stimuli only through the auditory segment of VA stimuli because the detection of auditory stimuli was too easy. So, the synchronously presented V stimuli did not improve the HRs to response the multimodal stimuli.

## 5 Conclusion

We investigated the influence of stimulus spatial locations on audiovisual interaction using a behavioural measure. As a result, the audiovisual interaction reduced ambiguity of stimuli and enhanced the behavioral response. Moreover, the behavioral enhancement of audiovisual stimuli depended on the spatial locations of audiovisual stimuli.

## Acknowledgment

A part of this study was financially supported by JSPS AA Science Platform Program and JSPS Grant-in-Aid for Scientific Research (B) (21404002).

## References

1. Calvert, G.A., Spence, C., Stein, B.E. (eds.): The Handbook of Multisensory Processing. MIT Press, Cambridge (2004)
2. Lippert, M., Logothetis, N.K., Kayser, C.: Improvement of visual contrast detection by a simultaneous sound. Brain research 1173, 102–109 (2007)
3. Gielen, S.C., Schmidt, R.A., Van den Heuvel, P.J.: On the nature of intersensory facilitation of reaction time. Percept. Psychophys 34, 161–168 (1983)
4. Hershenson, M.: Reaction time as a measure of intersensory facilitation. J. Exp. Psychol. 63, 289–293 (1962)
5. Posner, M.I., Nissen, M.J., Klein, R.M.: Visual dominance: an information-processing account of its origins and significance. Psychol. Rev. 83, 157–171 (1976)
6. Seitz, A.R., Kim, R., Shams, L.: Sound facilitates visual learning. Curr. Biol. 16, 1422–1427 (2006)
7. David Hairston, W., Hodges, D.A., Burdette, J.H., Wallace, M.T.: Auditory enhancement of visual temporal order judgmen. NcuroReport 17, 791–795 (2006)
8. Slutsky, D.A., Recanzone, G.H.: Temporal and spatial dependency of the ventriloquism effect. NeuroReport 12, 7–10 (2001)
9. Welch, R.B., Warren, D.H.: Immediate perceptual response to intersensory discrepancy. Psychol. Bull. 88, 638–667 (1980)
10. Morein-Zamir, S., Soto-Faraco, S., Kingstone, A.: Auditory capture of vision: Examining temporal ventriloquisum. Cognit. Brain Research 17, 154–163 (2003)
11. Spence, C., Driver, J.: Attracting attention to the illusory location of a sound reflexive crossmodal orienting and ventriloquism. NeuroReport 11, 2057–2061 (2000)
12. Raab, D.: Statistical facilitation of simple reaction times. Transactions of the New York Academy of Sciences 24, 574–590 (1962)
13. Eimer, M.: Attending to quadrants and ring-shaped regions: ERP effects of visual attention in different spatial selection tasks. Psychophysiology 36, 491–503 (1999)

14. Lewald, J., Getzmann, S.: Horizontal and vertical effects of eye-position on sound local-
    ization. Hear Res. 213, 99–106 (2006)
15. Simon, R., Sylvain, D., Pascal, D., Sabine, D.D., Jacques, H., Henrique, S.: Peripherally
    presented emotional scenes: A spatiotemporal analysis of early ERP response. Brain To-
    pogr 20, 216–223 (2008)
16. Anderson, R.: Aliasing in peripheral vision for counterphase gratings. J. Opt. Soc. Am A
    Opt. Image Sci. Vis. 13, 2288–2289 (1996)

# A Functional Model of Limbic System of Brain[*]

Takashi Kuremoto, Tomonori Ohta, Kunikazu Kobayashi, and Masanao Obayashi

Graduate School of Science and Engineering, Yamaguchi University,
755-8611 Tokiwadai 2-16-1, Ube, Japan
{wu,koba,m.obayas}@yamaguchi-u.ac.jp

**Abstract.** A functional model of limbic system of brain is proposed by combining four conventional models: a chaotic neural network (CNN), a multi- layered chaotic neural network (MCNN), a hippocampus-neocortex model and an emotional model of amygdala. The composite model can realize mutual association of multiple time series patterns and transform short-term memory to long-term memory. The simulation results showed the effectiveness of the proposed model, and this study suggests the possibility of the brain model construction by means of integration of different kinds of artificial neural networks.

## 1 Introduction

Artificial neural networks have been developed for decades from last century and successfully used in fields of function approximation, optimization, pattern recognition, intelligent control, and so on. Functionally, Hopfield network, a recurrent neural network in which all connections are symmetric, has the ability of stationary auto-association, meanwhile chaotic neural networks can realize dynamic association [1, 2]. Emotion models are also proposed and applied in control systems recently [3]. However, realization of an integrated artificial brain model including multiple functions of brain is still a high hurdle even though learning models, memory models, emotion models and other functional neural networks have been proposed.

The limbic system of mammalian brain locates on the under of brain including the parts of hippocampus, amygdala, anterior thalamic nuclei, and entorhinal cortex. It serves to a variety of functions including to transform short-term memory (working memory) into long-term memory (declarative memory), to control the emotional response and support decision of behaviors. In this paper, we propose a limbic system mathematical model, which combines several functional models, to present how input patterns are stored in hippocampus and transformed into long-term memory on cortex, and how emotion models participate in these processes. Our functional model of the limbic system is developed according to the following points:

1) Hippocampus plays an important role of transforming intermediate-term memory to long-term memory. Ito *et al* proposed a hippocampus-cortex model [4] and a hippocampus-neocortex model [5] which composed a circuit of cortex - dentate gyrus

---

[*] A part of this study was supported by JSPS-KAKENHI (No.18500230, No.20500277 and No.20500207).

N. Zhong et al. (Eds.): BI 2009, LNAI 5819, pp. 135–146, 2009.

– hippocampus – cortex to realize episode memory and long-term memory, respectively. For its characteristic structure of neuroanatomy and computational availability of mathematics, we adopt this hippocampus-cortex model as a part of our model of limbic system.

2) The structure of hippocampus is organized with stratified slices and chaotic response is observed appearing among these neurons. We proposed a new hippocampus-neocortex model to realize mutual association and long-term memory in our previous works [8, 9]. The main difference between our model and Ito *et al*'s model is that a model of CA3 layer of hippocampus was presented by multi-layered chaotic neural networks (MCNN) [7]. MCNN is also used to realize dynamical memory process in this study.

3) Amygdala plays important roles in emotional responses of brain, and neighbored to hippocampus in location. We consider that emotion can promote the efficiency of memory process, so adopt an emotion model proposed by Balkenius and Moren [3] [13] into our model of limbic system. The main components of the emotion model are thalamus, amygdala, orbitofrontal cortex and sensory cortex. The model has been used successfully in an intelligent control system [14] and we also used it to improve the chaos control process in hippocampus-neocortex model efficiently [10].

According to the knowledge of neuroanatomy, neurophysiology and physiological psychology, we propose a higher function model of limbic system fusing hippocampus-neocortex model, MCNN and emotion model described above in this paper.

## 2   A Model of Limbic System

A model of limbic system which is composed with a hippocampus-neocortex model [8, 9] and an amygdala model [3] [13] is shown in Fig. 1. The hippocampus-neocortex model consists of a memory circuit including cortex – dentate gyrus – hippocampus – cortex given by Ito *et al* [4] [5]. Entorhinal cortex and CA2 of hippocampus are omitted for their weak connections. A chaotic model of CA3 layer in hippocampus is adopted in with multi-layered chaotic neural networks (MCNN) [7] which combines two chaotic neural networks [1] instead of the single layer of Ito *et al*'s model to serve intermediate-memory processing. MCNN showed its effectiveness of mutual association of plural time series patterns and long-term memory formation in our previous simulations [7] [8] [9]. However, emotional effect is not considered in the memorization process of MCNN, likewise almost of conventional association systems. In fact, amygdala, which locates adjacent of hippocampus, plays important roles in the emotional responses such as fear and aggression, and also in the memorization process such as learning of Pavlovian fear conditioning and enhancing of long-term memory formation [14]. We adopt an amygdala model given by Balkenius and Moren [3] [13] here to evaluate and promote the efficiency of memory processing.

### 2.1   A Hippocampus-Neocortex Model

The dark dots and gray dots in Fig. 1 express neurons of neocortex and hippocampus, and symbols of CX1, CX2, DG, CA3 and CA1 denote the first layer of cortex, the

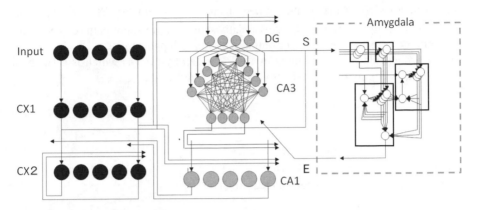

**Fig. 1.** A model of limbic system proposed in this paper. Neurons in cortex, hippocampus and amygdala are expressed with dark, gray and white dots respectively. Signals are expressed with arrow lines.

second layer of cortex, dentate gyrus, the third layer of hippocampus and the first layer of hippocampus respectively. The signal flow of the hippocampus-neocortex model is showed with arrows, i.e.: input stimuli (Input layer) → sensory memory (CX1) → short-term memory (CX2) → intermediate memory (DG) → Hebbian learning and chaotic processing of storage and recollection (CA3) → decoding (CA1) → long-term memory (CX2). The output of CA3, which is a result of chaotic memory processing, projects to Amygdala and the output of Amygdala is input to CA3 to realize chaotic state control of MCNN in CA3.

The dynamics of association cortex is given as following.

$$I_i = \begin{cases} 1 \cdots \text{excitatory} \\ 0 \cdots \text{inhibitory} \end{cases}. \tag{1}$$

$$x_i^{cx1}(t) = I_i(t). \tag{2}$$

$$x_i^{cx2}(t) = f\left(\sum_{j=0}^{N} w_{ij}^{cx2 \cdot cx2} x_j^{cx2}(t-1) + w^{cx2 \cdot cx1} x_i^{cx1}(t) + w^{cx2 \cdot ca1} x_i^{ca1}(t) - \theta^{cx}\right). \tag{3}$$

Where $I_i(t)$ is the output value of the $i$th neuron in the Input layer of association cortex, $x_i^{cx1}(t)$ and $x_i^{cx2}(t)$ are the output value of the $i$th neuron in CX1 and CX2 respectively, $w_{ij}^{cx2 \cdot cx2}$ denotes the weight of connection (variable) between the $j$th neuron (output) and the $i$th neuron in the CX2 layer, $w^{cx2 \cdot cx1}$ $w^{cx2 \cdot ca1}$ denote the weights of connections (fixed) between layers of CX1 and CX2, CX2 and CA1 respectively, $x_i^{ca1}(t)$ is the output of the $i$th neuron in CA1, $\theta^{cx}$ is a threshold value of cortex neurons, $N$ is the number of neurons in CX1 as same as CX2, $f$ expresses a step function. Eq. (3) endows CX2 associative function for input patterns and long-term memory formation function for output patterns of hippocampus.

The learning rule of the synapses in CX2 is given by Eq. (4) which is a Habbian rule using the different output of neurons in time $t$ and $t-1$.

$$\Delta w_{ij}^{cx2 \cdot cx2} = \alpha_{hc} \cdot x_i^{cx2}(t) x_j^{cx2}(t-1). \tag{4}$$

Where $\alpha_{hc}$ is a parameter of learning rate.

Hippocampus composes DG, MCNN and CA1 neurons. DG executes pattern encoding (Eq. (5)) with a competition learning (Eq. (6)).

$$x_i^{dg}(t) = \begin{cases} random \subset (0,1) & \cdots \ (initially) \\ f\left(\sum_{j=0}^{N} w_{ij}^{dg \cdot cx1} x_j^{cx1}(t) - \theta^{dg}\right) & \cdots \ (generally) \end{cases}. \tag{5}$$

$$\Delta w_{ij}^{dg \cdot cx1} = \beta_{hc} \cdot x_i^{dg}(t) x_j^{cx1}(t). \tag{6}$$

Where $w_{ij}^{dg \cdot cx1}$ denotes the weight of connection between the $i$ th neuron in CX1 (output) and the $j$ th neuron in DG, $x_j^{cx1}$ is the output of the $i$ th neuron in CX1, $\theta^{dg}$ is a threshold value of DG neurons, $\beta_{hc}$ is a learning rate and $\alpha_{hc} < \beta_{hc}$.

CA3 accepts the encoded information from DG and executes chaotic processing of storage and recollection with MCNN. It consists of two CNN layers which dynamics is given by Eq. (12) – Eq. (15) and one output layer which neuron's output is given by Eq. (7) – Eq. (9).

$$k = \arg\max_i \sum_{j=0}^{n} w_{ij}^{ca3out \cdot cnn1}(2x_j^{cnn1}(t) - 1). \tag{7}$$

$$k = \arg\max_i \sum_{j=0}^{n} w_{ij}^{ca3out \cdot cnn2}(2x_j^{cnn2}(t) - 1). \tag{8}$$

$$x_i^{ca3out}(t) = \begin{cases} 1 & \cdots \ (i = k) \\ 0 & \cdots \ (i \neq k) \end{cases}. \tag{9}$$

Here the $j$th neuron in CNN1 $x^{cnn1}{}_j(t)$ and the $j$th neuron in CNN2 $x^{cnn2}{}_j(t)$ are used to transform the output of MCNN by the $i$th neuron in output layer of MCNN $x_i^{c3out}(t)$. $w^{ca3out.cnn1}{}_{ij}$ and $w^{ca3out.cnn2}{}_{ij}$ denote the connections between the output layer and CNN1, CNN2 respectively. Learning rule is given as

$$\Delta w_{ij}^{ca3out \cdot cnn} = \frac{1}{n} x_i^{ca3out}(t) x_j^{cnn}(t).$$

Mutual associative memory process is realized by the output layer of MCNN receives the output of CNN1 and CNN2, i.e., Eq. (7) and Eq. (8), alternatively. In fact, when there are two time series patterns are processed by the two CNN layers of MCNN, one layer of CNN plays a role of static external stimuli meanwhile another layer of CNN executes chaotic storage or recollection processing dynamically. In another word, CNN layers fire alternatively in CA3. To switch their roles of CNNs, we used a simple threshold in [7] [8] [9], however, an emotional control given by subsection 2.2 and subsection 2.3 can raise the performance of MCNN [10].

CA1 in hippocampus decodes output pattern of MCNN which is expressed by $n$ neurons to patterns stored by associative cortex CX2 which has $N$ neurons (Eq. (10)). Hebbian learning rule is described by Eq. (11) where input from CX1 is a teacher signal ($w_{ij}^{ca1 \cdot cx1} = 1.0$).

$$x_i^{ca1}(t) = f\left(\sum_{j=0}^{n} w_{ij}^{ca1 \cdot ca3} x_j^{ca3out}(t-1) + w^{ca1 \cdot cx1} x_i^{cx1}(t) - \theta^{ca1}\right). \tag{10}$$

$$\Delta w_{ij}^{ca1 \cdot ca3} = \beta_{hc} \cdot x_i^{ca1}(t) x_j^{ca3out}(t). \tag{11}$$

Where $\beta_{hc}$ is a parameter of learning rate, $i = 1,2,...,N$.

CNN1 and CNN2 are Adachi & Aihara's CNN proposed in [1] [2] and combined to each other in our MCNN model [7]. The dynamics are described by Eq. (12) – Eq. (15).

$$x_i(t+1) = g(y_i(t+1) + z_i(t+1) + \gamma \cdot v_i(t+1)). \tag{12}$$

$$y_i(t+1) = k_r y_i(t) - \alpha x_i(t) + a_i. \tag{13}$$

$$z_i(t+1) = k_f z_i(t) + \sum_{j=1}^{n} w_{ij} x_j(t). \tag{14}$$

$$v_i(t+1) = k_e v_i(t) + \sum_{j=1}^{n} W_{ij}^* x_j^*(t). \tag{15}$$

Where $x_i(t)$ is the output value of $i$th neuron at time $t$, $n$ is the number of neurons of input layer, $w_{ij}$ is the weight of connection from $j$th neuron to $i$th neuron, $y_i(t)$ denotes internal state of $i$th neuron, $\alpha$ is a learning rate of $i$th neuron, $k_f$, $k_r$, $k_e$ are damping rates, $a_i$ is a parameter as the summation of threshold and external input, $\gamma$ is a rate of effectiveness from another layer, $W^*_{ij}$ denotes the weight of connection from $j$th neuron of different CNN layer to $i$th neuron in current CNN layer, $x^*_j$ $(t)$ is the output value of $j$th neuron in different CNN layer at time $t$. $g(\cdot)$ is a sigmoid function.

When a MCNN is composed by two CNN layers, learning rules of the connections between the CNN layers $W^*_{ij}$ are described by Eq. (16) and Eq. (17).

$$\Delta W_{ij}^{cnn1\cdot cnn2} = \beta \cdot x_i^{cnn1}(t)x_j^{cnn2}(t).$$ (16)

$$\Delta W_{ij}^{cnn2\cdot cnn1} = \beta \cdot x_i^{cnn2}(t)x_j^{cnn1}(t).$$ (17)

Where $\beta$ is a parameter of learning rate, usually $\beta = 1/m$, $m$ is the number of stored patterns.

Conventionally, we calculated the temporal change of the internal state of a CNN layer $\triangle x(t)$, and when $\triangle x(t)$ is less than a threshold $\theta$, the chaotic retrieval of the layer is stopped by changing values of parameters $k_r$, $k_f$ into zero, as a result, the CNN layer becomes to a Hopfield model. The recalled pattern of one CNN layer provides an input pattern to the other CNN layer, and the network realizes mutual association and one-to-many retrieval for plural time series patterns [7] [8] [9].

## 2.2 An Amygdala-Hippocampus Model

Balkenius & Moren's computational amygdala model [3] [13], which is shown in the right part of limbic system model in Fig. 1, is combined with hippocampus-neocortex described above to evaluate and promote the performance of memory processing. Recently, we proposed an amygdala-hippocampus model which showed faster storage ability and higher precision of recollection of plural time series patterns dynamical association [10]. The main idea of the adoption of emotional model comes from the consideration that unstable state of hippocampus may result emotional response, i.e., arousal of amygdala, or the high value of amygdale output may enhance memory processing happened in hippocampus.

The dynamics of the amygdale model is described as follows:

$$A_i = V_i S_i.$$ (18)

$$O_i = W_i S_i.$$ (19)

$$E = \sum_i A_i - \sum_i O_i.$$ (20)

$$\Delta V_i = \alpha_{AMY}(S_i \max(0, R - \sum_j A_j).$$ (21)

$$\Delta W_i = \beta_{AMY}(S_i \sum_j (O_j - R)).$$ (22)

Where $S_i$ denotes input stimuli from sensory cortex and thalamus to $i$th neuron in amygdala, $A_i$ denotes the output of $i$th neuron in amygdala, $O_i$ denotes the output of $i$th neuron in orbitofrontal cortex, $E$ is the output of amygdala which control the internal state of MCNN by tuning damping parameters as same as our conventional method. $\alpha_{AMY}, \beta_{AMY}$ are learning rates, $V_i, W_i$ are variables of connections in Amygdala model. Reward $R$ coming from other area of brain is used to renew Eq. (21) and Eq. (22) which belongs to reinforcement learning algorithm.

Suppose that the unstable degree of a CNN provides a reward $R$ to amygdala model, then an emotional control can be realized by Eq. (18) – Eq. (27).

$$S_i = \frac{\sum_{j=i*\Delta N}^{\Delta N*(i+1)}(x_j(t+1) - x_j(t))}{\Delta N},\tag{23}$$

where $i = 0,1,\cdots,N_{AMY} - 1$.

$$\Delta N = \frac{n}{N_{AMY}}.\tag{24}$$

$$R = \begin{cases} 1.0 & \cdots & (F > n_R) \\ 0.0 & \cdots & (else) \end{cases}.\tag{25}$$

$$F = \frac{\sum_{i=0}^{N_{AMY}} g(S_i - \theta_R)}{N_{AMY}}.\tag{26}$$

$$CNN\ state : \begin{cases} chaotic & \cdots & (E > \theta_{AMY}) \\ non-chaotic & \cdots & (else) \end{cases}.\tag{27}$$

Where $S_i$ in Eq. (23) corresponds to input of Amygdala (Eq. (19) and Eq. (20)), $x_j$ denotes the output of chaotic neuron in CNN layers, $N_{AMY}$ is the number of amygdala layer in Amygdala model, n is the number of CNN layers, $R$ is a reward according to Eq. (25) where $F$ expresses fire rate of the neurons in amygdala layer, $\theta_R$ is a threshold of the output of the amygdala layer, $n_R$ is a threshold of the reward function Eq. (25), $g(\cdot)$ is a Sigmoid function. The output of Amygadala $E$ controls state of MCNN with threshold $\theta_{AMY}$ which is described in Eq. (27).

## 3  Simulations

To confirm the effectiveness of the proposed model of limbic system, we performed two kinds of simulations using a personal computer loaded a Pentium 4 CPU. The fist simulation was designed to compare the mutual association abilities between conventional MCNN and Amygdala – Hippocampus model. This simulation was also described in [10], however, new statistical results would be reported more in this paper. The second simulation was designed to compare one-to-many association abilities, and long-term memory formation abilities between Amydala-Hippocampus model and the model of limbic system proposed in this paper.

### 3.1 Simulation of Mutual Association

Two time series patterns used in mutual association simulation of MCNN and Amygdala-Hippocampus model are as same as those in [7] and [10]. All parameters value or their initial value used in these simulations were decided by empirical knowledge or according to the previous works and they are shown in Table 1. The comparison of storage time and recollection time of different models is shown by Table 2. Amygdala-Hippocampus model had a faster storage than MCNN however a slower recollection. The results can be explained as that emotional model enhanced the storage processing in the meaning of efficient and enhanced the recollection more "carefully" in the meaning of precision. In fact we confirmed that retrieved patterns using emotional control method showed their completeness meanwhile MCNN failed sometime. Illustrations of the results by MCNN are omitted here for want of space, and results by Amygadala-Hippocampus model are shown in Fig. 2.

### 3.2 Simulation of Long-Term Memory

Two time series patterns shown in Input layer of Fig. 3 similar to the simulations of Ito *et al*'s [5] were used to investigate the association and long-term memory formation abilities of proposed model of limbic system. Binary patterns in each time series were orthogonal, and a 4-step interval between the two time series was set to distinguish them. The procedure of simulation is described according to time sequence as following:

A. Input the two time series patterns. The first pattern of each time series was same to serve as a key pattern of one-to-many association.

B. Intermediate memory recollection. Input the key pattern to associate time series patterns stored on the time A.

C. Recollection without hippocampus. Stop the output of hippocampus and amygdala temporarily to investigate long-term memory formation on CX2.

D. Consolidate long-term memory with a long-term potentiation (LPT) process, i.e., input the key pattern repeatedly. Hippocampus works to transform intermediate memory to long-term memory.

E. Recollection of long-term memory. Stop hippocampus and amygdala to investigate long-term memory formation on CX2.

One simulation result of one-to-many association and long-term memory formation are shown in Fig. 3. One time series pattern was stored in CX2, while the rate of storage including failed one is reported by Table 3. The output of amygdala model during storage and recollection processes is shown in Fig. 4. Comparing with conventional hippocampus-neocortex model [8] [9], the model of limbic system proposed raised 8% rate of successful recollection.

**Table 1.** Parameter values (or initial values) used in the simulations

| Symbol | Description | Value |
|---|---|---|
| $N$ | Number of neurons in association cortex | 50 |
| $n$ | Number of neurons in association cortex | 30 |
| $w_{ij}^{cx2\cdot cx1}$ | Weight of connection from CX1 to CX2 | 1.0 |
| $w_{ij}^{cx2\cdot ca1}$ | Weight of connection from CX2 to CA1 | 1.0 |
| $\alpha_{hc}$ | Learning rate in association cortex | 0.0015 |
| $\beta_{hc}$ | Learning rate in hippocampus | 1.0 |
| $\alpha_{AMY}$ | Learning rate in amygdala part | 0.2 |
| $\beta_{AMY}$ | Learning rate in ortbitofrontal part | 0.8 |
| $k_f$ | Damping coefficient in CNN | 0.02 |
| $k_e$ | Damping coefficient in CNN | 0.02 |
| $k_r$ | Damping coefficient in CNN | 0.1, 0.9 |
| $N_{AMY}$ | Number of neurons in Amygdala | 10 |
| $\theta$ | Threshold of conventional control | 5 |
| $\theta_{AMY}$ | Threshold of emotional control | 0.0-1.0 |
| $\theta_R$ | Threshold of Amygdala neurons | 0.3, 0.6, 0.9 |
| $\theta^{cx}$ | Threshold of association cortex | 0.5 |
| $n_R$ | Threshold of Amygdala reward | 0.0-1.0 |

**Table 2.** Comparison of memory processing performance with results of mutual association simulations

| Model | Storage time (steps) | Recollection time (steps) |
|---|---|---|
| Conventional MCNN [7] [8] [9] | 18 | 17 |
| Amygdala – Hippocampus model | 8 | 35 |

**Table 3.** Comparison of successful rate of recollection with results of one-to-many times series pattern association simulations (%)

| Models | Time series A | Time series B | Failed |
|---|---|---|---|
| Hippocampus – neocortex model [8] [9] | 7 | 3 | 90 |
| Proposed model of limbic system | 9 | 9 | 82 |

CNN1  CNN2

store

recollect

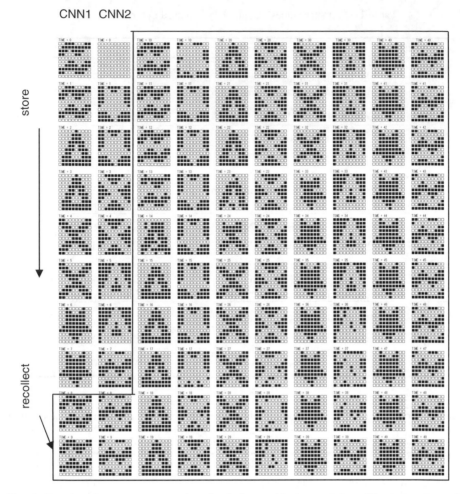

**Fig. 2.** Results of memory processing simulations with Amygadala – Hippocampus model proposed in [10]

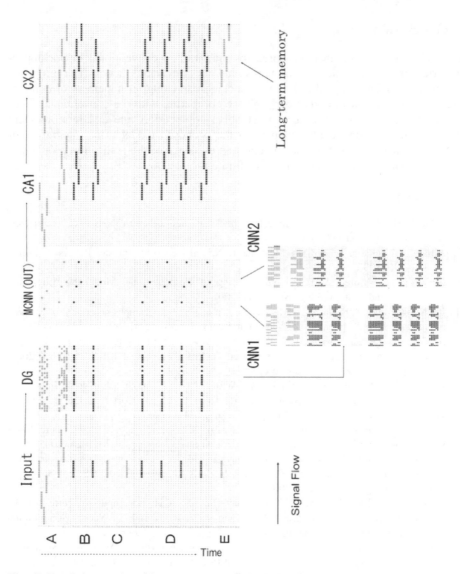

**Fig. 3.** Simulation results of long-term memory formation. Two time series patterns were presented during learning process and a piece of a common pattern was used to be a clue pattern until long-term memory was formed on CX2 layer.

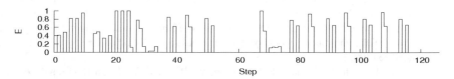

**Fig. 4.** The change of output of Amygdala during storage and recollection process

# 4 Conclusion

A model of limbic system is proposed by combining a conventional hippocampus-neocortex model, chaotic neural works and an amygdala model. The proposed mathematical model can realize mutual association and long-term memory of multiple time series patterns with higher performance comparing with the conventional models. Different learning rules, such as Hebbian competition rule and reinforcement learning rule, are functionally adopted in the proposed model. This integration of memory models and emotion models gives an evidence of the realization probability of the computational artificial brain in the future.

# References

1. Adachi, M., Aihara, K.: Associative Dynamics in Chaotic Neural Network. Neural Networks (10), 83–98 (1997)
2. Aihara, K., Takabe, T., Toyoda, M.: Chaotic Neural Networks. Physics Letters A. 144(6-7), 333–340 (1990)
3. Balkenius, C., Morén, J.: Emotional Learning: A Computational Model the Amygdala. Cybernetics and Systems 32(6), 611–636 (2000)
4. Ito, M., Kuroiwa, J., Sawada, Y., Miyake, S.: A model of hippocampus-neocortex for episodic memory. In: Proc. 5$^{th}$ Intern. Conf. Neural Information Processing, 1P-16P, pp. 431–434 (1998)
5. Ito, M., Miyake, S., Inawashiro, S., Kuroiwa, J., Sawada, Y.: Long term memory of temporal patterns in a hippocampus-cortex model. Technical Report of IEICE. In: NLP 2000, Vol.18, pp. 25–32 (2000) (in Japanese)
6. Kajiwara, R., Takashima, I., Mimura, Y., Witter, M.P., Iijima, T.: Amygdala Input Promotes Spread of Excitatory Neural Activity from Perirhinal Cortex to the Entorhinal-Hippocampal circuit. J.europhysiology 89, 2176–2184 (2003)
7. Kuremoto, T., Eto, T., Kobayashi, K., Obayashi, M.: A multilayered chaotic neural network for associative memory. In: Proc. of SICE Annual Conf., pp. 1020–1023 (2005)
8. Kuremoto, T., Eto, T., Kobayashi, K., Obayashi, M.: A chaotic model of hippocampus-neocortex. In: Wang, L., Chen, K., Ong, Y.S., (eds.) ICNC 2005. LNCS, vol. 3610, pp. 439–448. Springer, Heidelberg (2005)
9. Kuremoto, T., Eto, T., Kobayashi, K., Obayashi, M.: A Hippocampus-Neocortex Model for Chaotic Association. In: Chen, K., Wang, L. (eds.) Trends in Neural Computation (Studies in Computational Intelligence), vol. 35, pp. 111–133 (2006)
10. Kuremoto, T., Ohta, T., Obayashi, M., Kobayashi, K.: A Dynamic Associative System by Adopting an Amygdala Model. Artif. Life and Robotics 13(2), 478–482 (2009)
11. MaGaugh, J.L., Cahill, L., Roozendaal, B.: Involvement of the Amygdala in Memory Storage: Interation with Other Brain Systems. In: Proc. Natl. Acad. Sci. USA, vol. 93, pp. 13508–13514 (1996)
12. Mizutani, S., Sano, T., Uchiyama, T., Sonehara, N.: Controlling Chaos in Chaotic Neural Networks. Electronics and Communications in Japan, Part III: Fundamental Electronics Science 81(8), 73–82 (1998)
13. Morén, J., Balkenius, C.: A Computational Model of Emotional Learning in the Amygdala. In: Proc. of the 6th Intern. Conf. on the Simulation of Adaptive Behavior. MIT Press, Cambridge (2000)
14. Rouhani, H., Jalili, M., Araabi, B.N., Eppler, W., Luscas, W.: Brain Emotional Learning Based Intelligent Controller Applied to Neurofuzzy Model of Micro-Heat Exchanger. Expert Sys. with Appli. 32, 911–918 (2007)

# Data Explosion, Data Nature and Dataology

Yangyong Zhu[1], Ning Zhong[2], and Yun Xiong[1]

[1] School of Computer Science, Fudan University
Shanghai 200433, P.R. China
yyzhu@fudan.edu.cn, yunx@fudan.edu.cn
[2] Dept. of Life Science and Informatics, Maebashi Institute of Technology
Maebashi-City 371-0816, Japan
zhong@maebashi-it.ac.jp

**Abstract.** The essence of computer applications is to store things in the real world into computer systems in the form of data, i.e., it is a process of producing data. Some data are the records related to culture and society, and others are the descriptions of phenomena of universe and life. The large scale of data is rapidly generated and stored in computer systems, which is called *data explosion*. Data explosion forms *data nature* in computer systems. To explore data nature, new theories and methods are required. In this paper, we present the concept of data nature and introduce the problems arising from data nature, and then we define a new discipline named *dataology* (also called *data science* or *science of data*), which is an umbrella of theories, methods and technologies for studying data nature. The research issues and framework of *dataology* are proposed.

## 1 Introduction

According to the recent IDC research report entitled "As the Economy Contracts, the Digital Universe Expands" [1], the amount of new digital information reached about 486 billion gigabytes in 2008 and increased 3 percent faster than IDC previous projection. The digital universe is expected to be double in size every 18 months. In 2012, five times as much digital information will be generated versus 2008. When the data are explosively increasing, they also become more complicated and diversified. At the *IBM Information on Demand 2009* conference, experts pointed out that in the world almost 15 GB (gigabytes) data are produced every day. These data come from various equipments including sensors, RFIDs, meters, GPSs, etc., and at least 80 percent of new data are unstructured, such as Web contents, Web logs, email, image, video, audio, and so on.

All the facts mentioned above indicate that data explosion has happened and been spreading. In fact, *data explosion* is the course that data in computer systems explosively increase since human continuously stores data when using the computers. During the course of data explosion, the mass data appear multiple natural features including out of control, unknown, diversity and complexity. Therefore, data explosion forms *data nature*. Studying data nature is an effective

N. Zhong et al. (Eds.): BI 2009, LNAI 5819, pp. 147–158, 2009.

approach to study real nature, for example, the researches on *Bioinformatics* [2], *Brain Informatics* [3] and *Behavior Informatics* [4] indicate that we can study life and human brain by studying related scientific data.

This paper proposes *dataology* (also called *data science* or *science of data*) which is an umbrella of theories, methods and technologies for studying data nature. The rest of this paper is organized as follows. Section 2 introduces data explosion. Section 3 describes data nature including natural features and evolution, as well as key issues in data nature. Section 4 presents dataology as a new research discipline and provides its framework and content. Finally, Section 5 gives concluding remarks.

## 2   Data Explosion

Data are increasing explosively with the development of human being. Trying to remember is the instinct of human being. From time immemorial, using brain to remember experienced things is the primary means. Because of some unknown reasons, human memory cannot retain everything they read. The memories in human brain are also unreliable. Thus, human seeks various tools to help them to memorize all along. Originally, human carved figures and characters on hard objects to assist remembering. They found that the information recorded out of brain was convenient for transmission and communication, therefore the human instinct of recording information is deepened.

The inventions of papermaking and printing brought about the first data explosion [1] during which a mass of natural things (including natural phenomena, culture, society, etc.) are represented by characters or figures, and then printed into books or materials. That is, the information about a period of history can be recorded into a book for memorizing and transmitting, such as the Bible, the Records of the Grand Historian, and so on. The information can be stored for a long time, replicated many times, and spread widely. During the course of this data explosion, the authors/publishers produced information, and the books/libraries stored and conveyed information.

The inventions of computers (especially Internet/WWW) and storage devices brought about the second data explosion. It is a process that data in computer systems explosively increase because human continuously stores data when they use the computers. During this explosion, all of books in the earlier libraries and previous publications (i.e., main productions in the first data explosion) can be stored into a personal computer, even a removable hard disk.

So far no proof indicates that there exists a kind of device which can replace computers and storage devices. In the future, a certain kind of man-made being

---

[1] The term "information explosion" and "data explosion" are usually replaceable. Information is the explanation of data, and data is the symbolic representation of information. Therefore, in general, the more the information, the more the data required to be stored is. Conversely, the more the data, the more the information can be expressed is. In this paper, we use the term "data explosion".

would be created, who has the camera memory and its memory retains everything it reads. It would exhibit powerful processing ability. This will bring the next data explosion.

Data explosion makes people lost in the mass data, which can be illustrated in the following aspects:

- Difficult to ensure the truth of data: we cannot know if the data obtained from computer systems (such as Internet/WWW) are true. This will lead that we do not know which are usable though the mass data we own.
- Difficult to share data: data sharing becomes more and more difficult though it is one of goals of computer systems to provide the ability of data sharing. A great amount of data are produced every day, we do not know what need to be shared, as well as how to share.
- Difficult to keep data consistent: we cannot ensure the consistency of data. For example, we often get the different results when we query the same object on different websites.

Though we are continuously developing new technologies, such as grid computing [5], cloud computing [6]), and wisdom-Web computing [7], the above problems become more and more serious.

## 3   Data Nature

### 3.1   What Is Data Nature?

The development of the world is the course in which human being continuously explores real nature (universe and life) and builds up culture and society. When the human activities and their results are stored into computer systems, we create a data nature unconsciously.

During data explosion, more and more data are stored into computer systems. They have various categories and formats. For example, there exist the following data categories:

- Personal data: The personal data are often stored in personal computers, including personal privacy data and work data. The personal work data involve various contents, such as the the data for the company/organization, the data for the job, and so on. The personal data are also dispersed in the Internet, which are often neglected though it is a kind of important personal data.
- Enterprise data: The enterprise data are stored in enterprise computer systems. They are respectively from enterprise operations, clients, competitors, trades, etc.
- Government data: The government data are stored in governmental computer systems, which include government operation data, social resource data, economics of population data, etc.
- Public data: The public data are stored in public websites, which can be accessed by Web search engines.

- Data in various kinds of languages: Different nationalities use different languages, such as English, Chinese, Arabic, and so on. Therefore, the data in different languages have been produced.
- Geographical data: The geographical data describe the geographical situations and changes in our countries or areas, which include the data related to space, ocean, earth, country, etc.
- Life data: The life data implicate a great deal of life characteristics and information, which include DNA sequence data, protein sequence data, and cognitive and medical brain data, etc.
- Cultural and social data: The cultural and social data describe the development of the human being and the society, which include human behaviors, economic data, etc.
- Internet data: The Internet data are dispersed in the Internet. They are easy to access, but contain numerous garbage and viruses. The appearance of Internet data makes the data in computer systems to show more natural features.

The formats of the data in data nature include:

- Special format data: The special format data are the data produced by professional digital equipments, such as medical image data (including x-ray, MRI, CT, EEG, etc.), cognitive brain image data, GIS data, multimedia data, which can be collected and processed by professional equipments or software.
- General format data: The general format data are the data stored in general format. In the early stage of computer applications, most of data are stored in relational databases and managed by general database management systems, such as Oracle or DB2. These data have clear structures and are easy to process.

The mass unknown data in computer systems are the basis of data nature. We do not know if the data from the Internet are true; we query the same object on different websites, but often get different results; maybe in the Internet a database has indicated that human being will face energy crisis, but we cannot grasp this knowledge; we input DNA sequences into computers, but we do not know what they indicate? what law they have? which fragments of DNA make the differences among people? how genes change during the evolution of species? whether gene evolution or mutation exists? and so on.

Though human being have produced data and are continuously producing data, these data have shown various natural features as follows:

- Out of control: The explosive increase of data leads that human cannot completely control it. Human also cannot control the appearance and spread of computer viruses, the deluge of SPAMs, the jam of NII (National Information Infrastructure) [8] due to network attack, etc.
- Unknown: Since HGP (Human Genome Project) [9] launched, a great amount of DNA data are stored into computers. However, people do not know what

these DNA sequences indicate, as well as how genes change during the evolution.

- Diversity: As mentioned above, there are various data categories, they are from space, ocean, biology, brain, multiple languages, various trades, as well as in the Internet/out of the Internet, public/private, therefore, the data in data nature exhibit diversity.
- Complexity: The mass data in computer systems are complex, they have various data formats and there exist many associations and relationships among these data.

Thus, natural features of data are more and more obvious. Various data forms and data "regions/areas" or data "tribes" have come into being. Though during data explosion we cannot distinctly describe the data nature, in fact, we have already worked and lived in it. The Digital Earth project [10] is in process. Through it, we are gradually transforming real nature into data nature.

### 3.2  Key Issues in Data Nature

As shown in Fig. 1, our culture and society are built on real nature at the beginning, and then the computer science and technology help people store both "culture and society" and "real nature" into computer systems when the computer was invented (as shown in Fig. 2). This meets both practical requirements and the human instinct of remembering.

As shown in Fig. 3, culture and society will be built on both real nature and data nature, and supported by computer science and technology. It means that our culture and society will rely increasingly on data nature.

In data nature, we will face many new problems. For example, if one asks how many papers are related to DNA in *science*, it is easy to answer; if he asks how

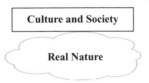

**Fig. 1.** Culture and society are built on real nature

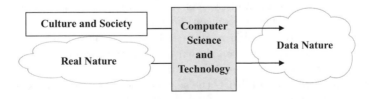

**Fig. 2.** From real nature, culture and society to data nature

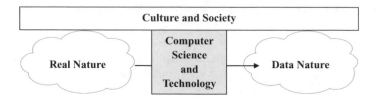

**Fig. 3.** The culture and society based on both real nature and data nature

many fields are studied in *science*, it is difficult to answer; if he asks which paper is associated with each other in *science* or what is indicated by all the papers in *science*, it is more difficult to answer. Therefore, we say that *science* itself is a valuable research topic. *Science* has been digitized and stored into computer systems in the form of data. It forms a "data tribe" or "region" like human tribe or region in real nature. Similarly, a library, a DNA database and a DOD (Department of Defense) system are bigger data tribes or regions which need us to explore and study.

In fact, we will face the following problems in data nature:

− How to recognize data nature?
− How to recognize real nature through data nature?
− How to live in both data nature and real nature?
− How to recognize whether the contents of data nature truly represent the real nature?
− How to develop and utilize data nature for the part superior to real nature?

Human existing foundation is material, that is, all the basic necessities of life, such as clothing, food, shelter and transportation, are based on materials. Computer systems have changed the foundation, for example, life habit, thinking, standpoint, morality, law, etc., have been changed. People rely increasingly on data in computers rather than the materials. Now human lives have been occupied by more and more data.

## 4   Dataology

### 4.1   What Is Dataology?

As mentioned above, the research achievements on real nature are stored into computer systems in the form of data which forms data nature. The exploration on data nature will be on a higher level than before. Many principles and laws in real nature, such as *prime number, fibanacci sequence, golden ratio, pareto principle*, etc, are also available in data nature (e.g., DNA data tribe). Many nice things in the world have been shown to us in the form of data, so now it is necessary for us to discover them from data.

With the forming of data nature, the science and technology of exploring data become more and more important. One significant discipline called *dataology* is forming, which takes data as a research object.

For a long time, the researches and technologies aiming at data mainly focus on data storage and management, such as database technology. Their primary goal is to model real nature, rather than to take data themselves as a research object.

The appearance of data mining technology [11] means that people began to study the laws aiming at data in computer systems. In the field of Internet, more and more researches focus on network behavior, network community, network search, and network culture. Because of the accumulation of data, newly disciplines, such as bioinformatics and brain informatics, are also typical dataology centric research areas. For instance, DNA data in bioinformatics are the data that describe natural structures of life, based on which we can study life using computers.

Brain informatics, which emphasizes a *systematic* approach to investigating human information processing mechanisms, including measuring, collecting, modeling, transforming, managing, mining, interpreting, and explaining multiple forms of brain data obtained from various cognitive experiments by using powerful equipments, such as fMRI (functional magnetic resonance imaging) and EEG (electroencephalogram) [3]. A tangible goal of brain informatics studies is to form a brain data area with a conceptual brain data model namely *Data-Brain*, which represents functional relationships among multiple human brain data sources, with respect to all major aspects and capabilities of human information processing systems (HIPS), for systematic investigation and understanding of human intelligence. In fact, Data-Brain construction is to form brain data area in computer. On one hand, developing such a Data-Brain is a core research issue of brain informatics. Systematic brain informatics research needs a Data-Brain to describe related knowledge and to annotate various human brain data sources, in order to support data sharing and data integration. Based on this way, it provides a long term, holistic vision to uncover the principles and mechanisms of underlying HIPS. On the other hand, the brain informatics methodology supports such a Data-Brain construction. The Data-Brain is used to model various human brain data related knowledge, involving data themselves, data production and data disposal.

Thus, we need a special discipline *dataology* and pay much attention to studying data themselves. Dataology is an umbrella of theories, methods and technologies for studying data nature. It will provide basic theories and methods for many (maybe all) disciplines and fields, including *data acquisition, data analysis, data awareness, data management*, etc. These basic theories and methods will be applied on various fields for developing special theories, methods and technologies, which will form dataology for specific domains.

The main research issues of dataology are:

- To form data areas in various domains;
    Massive datasets have driven research, applications, and tool development in business, science, government, and academia. The continued growth in data collection in all of these areas ensures the fundamental problem of dataology addresses, namely how the component (i.e., data area) of data nature is

formed in various domains. We are experiencing a strong demand for more powerful, intelligent computing paradigms for large-scale data measuring, collecting, modeling, transforming, exploring and managing [7].

- To study the structures of datasets in data nature;
  In data nature, a dataset which we will deal with may be a hard disk full of data, or a database managed by a certain DBMS, or a Web server, therefore, we need to study how to access these data from these media or circumstances. It is critical to analyze storage structure and logical structure of a dataset, which is called the study on the structures of a dataset.
- To acquire available data from data nature;
  Like to explore the golden or the oil, we will acquire available data from data nature. However, it is different from data mining because it only gets original data rather than process and analyze data. In general, available data should be acquired from various data sources in data nature and these data should be integrated. Sometimes they should be stored in the data warehouse after data cleaning.
- To prove the rules of data nature by theoretical methods;
  Like the research on science, we need to establish many theories and methods, and present hypothesis, induction, deduction and inference, and build up logical and theoretical systems in order to solve various problems generating from data nature.
- To discover the rules of data nature by experimental means;
  There are a lot of propositions and rules to be verified. Furthermore, many valuable results will produce through experiments, which is similar to the chemical experiments. Therefore, we need to establish various experimental systems and means in order to discover the rules in data nature.
- To develop and utilize data resources in data nature.
  Developing and utilizing data resources is a goal to research data nature, which will support the research on natural science and social science and serve for human life and social development. We believe that data resources are the most important resources in this century (perhaps more important than oil and coal), therefore, it is an important issue to develop and utilize data resources in data nature, which is an important topic in dataology.

## 4.2   The Framework of Dataology

The framework of dataology is shown in Fig. 4. This framework includes two main parts: foundations of dataology and applications of dataology (e.g., universal dataology, life dataology, behavior dataology, etc).

**Foundations of Dataology.** Foundations of dataology is composed of three aspects: data acquisition, data analysis and data awareness. They can be divided in more detail (see Fig. 4). All these technologies require data management. There are some existing technologies including data integration, data management (e.g., file system, database management system and data warehouse, etc.), data mining, data visualization, etc. They are developing continuously.

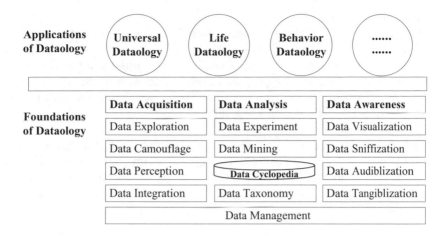

**Fig. 4.** The framework of dataology

In addition, dataology needs to develop more new technologies:

– **Data Experiment**

Most of people consider that bioinformatics makes biological experiment to become data computing. However, we believe that bioinformatics is more like biological data experiment. For example, we have a gene sequence of SARS virus in which "A", "C", "G" and "T" are represented as the points with different colors. When we optionally change sequences or duplicate parts of sequences (this just like to try to mix multiple reagents in chemical experiments), we may find an "S" picture. If that is quite true, the result is very valuable. Such cases also exist in other fields (e.g., brain informatics). This means that data experiment technology will be on demand.

Data experiment is to use various known or unknown methods to deal with a dataset in order to discover special features and laws. It focuses on the randomness of methods and the unpredictability of results. This is different from data mining in which the selection of methods is based on the prospective results.

– **Data Camouflage and Perception**

The data camouflage is to camouflage the private data which are exposed in the public. Different from data security (privacy protection, privacy mining), data camouflage focuses it efforts on camouflaging the data in the public, rather than storing data in a safe place to prevent invasion.

In the field of computer, a logion says, "garbage in garbage out". However, current problems are that we do not know which data are garbages, as well as which data are the disguise of valuable data. How to obtain the valuable data under this circumstance? This involves data perception which is to percept the camouflaged data. It can be regarded as the reverse of data camouflage.

- **Data Taxonomy**

  Data taxonomy is to classify data to form the pedigree of data and history of development. Because of the forming of data nature, the data exhibit diversity and have various categories. The data in data nature need to be classified according to data purpose or relationships, which is similar to classify the things in real nature according to species, history or culture. This new technology is called data taxonomy.

- **Data Awareness**

  The sense of human being includes vision, audition, sniff, touch, etc. Thus, if people want to feel data nature as feeling real nature, only data visualization technology is not enough. Dataology needs to develop new technologies of data awareness including data sniffization, data audibization, data tangiblization, etc.

**Applications of Dataology.** Because of the forming of data nature, people are unable and unnecessary to use all of data and pursue data consistency. People will use data in part under their lives and work circumstance. Thus, for obtaining the better results, new data technologies will be the special technologies aiming at different fields and circumstances instead of the general technologies. For example, the technologies of bioinformatics are just the special technologies. Therefore, specific domain dataology will create corresponding data technologies.

The applications of dataology consist of universal dataology, life dataology, behavior dataology and so on, in which, some specific domain dataologies have been formed from the viewpoint of dataology, because they all work on data nature. For example, we believe that both bioinformatics and brain informatics should belong to life dataology, and behavior informatics should belong to behavior dataology. In addition, various dataologies will come into being, such as spatial dataology, oceanic dataology, financial dataology, and so on.

Let us take *brain informatics* also called *brain dataology* as an example to illustrate the applications of dataology in the framework. From the viewpoint of dataology, the key steps in brain dataology research are shown in the followings:

1. Brain activities experiments:

   To do the experiments aiming at brain activities using some devices, such as fMRI, OT, ERP/EEG, etc, and collect the experimental data which will be stored into the data nature in the computer systems.

2. Brain data collection:

   To collect the brain data from data nature, for example, we can collect the brain data through the experiments, the literatures or the public databases.

3. Brain data integration:

   To integrate various types of brain data and build the brain data warehouses according to the demands on the brain science research in order to provide the comprehensive data for systematical research.

4. Brain data mining:

   To discover the rules in human brain activities when people participate the cognition activities, such as, solving problems, reasoning, making decisions, learning, etc. In general, new data mining algorithms should be developed according to the demands on brain science research.

5. Mining results analysis:

   To analyze the brain activities rules and cognition approaches from the results of data mining, because the mining results indicate the rules in the brain activity data. In this stage, brain scientists, psychologists and dataologists are expected to collaborate.

6. Mining results verification:

   To do the experiments and verify the results from brain data mining (i.e., the potential rules of brain activities). If it is true, the knowledge of brain activities will be accepted, conversely, the brain activities experiments in Step 1 should be improved, for example, designing the newly instruments and devices.

7. Brain activities experiments improvement:

   To improve the brain activities experiments according to the verified results in Step 6. And so on, over and over again, human recognition ability on the brain is improved gradually.

In the above steps, the tasks in Steps 1, 6 and 7 are administrated by brain scientists, or brain scientists are expected to participate, whereas, the tasks in Steps 2, 3, 4 and 5 are performed by dataologists.

It can be expected that all of the existing fields can form the corresponding dataology. Thus, we say that dataology is one of the most important disciplines in the $21^{st}$ century.

## 5   Conclusion

We explore real nature and use computers to represent our discoveries, society, nature and human being. Data have been produced explosively and a complex data nature has been created unconsciously. The history of human being and society will become the history of data. Therefore, it is necessary for human to research data nature, and new theories and methods are required, which is the goal to present the new discipline *"dataology"*.

Data nature (i.e., data in computer systems) is the research object of *dataology*. There will be many new research contents and methods, such as data experiment, data taxonomy, data awareness, and so on. In the $21^{st}$ century, people will live in both real nature and data nature. Dataology as an emerging discipline will become one of the most important disciplines. The data technology also will become one of the most important technologies in the future.

## Acknowledgement

The research was supported in part by Shanghai Leading Academic Discipline Project under Grant No. B114. The authors would like to thank professor Yixue Li, professor Yang Zhong and professor Longbing Cao for their suggestions. And we would like to thank Jianhui Chen and Xue Bai for collecting materials.

# References

1. http://www.emc.com/collateral/demos/microsites/idc-digital-universe/
   iview.htm (accessed May 2009)
2. Krawetz, S., Misener, S. (eds.): Bioinformatics Methods and Protocols. Humana
   Press (2000)
3. Zhong, N., Liu, J., Yao, Y.Y., Wu, J., Lu, S., Li, K. (eds.): Web Intelligence Meets
   Brain Informatics. LNCS (LNAI), vol. 4845, pp. 1–31. Springer, Heidelberg (2007)
4. Cao, L.B.: Behavior Informatics and Analytics: Let Behavior Talk. In: Proceedings
   of the 2008 IEEE International Conference on Data Mining Workshops (2008)
5. Berman, F., Fox, G., Hey, A. (eds.): Grid Computing: Making the Global Infrastructure a Reality. John Wiley & Sons, Chichester (2003)
6. Hayes, B.: Cloud Computing. Communications of the ACM 51(7), 9–11 (2008)
7. Zhong, N., Liu, J., Yao, Y.Y.: Envisioning Intelligent Information Technologies
   through the Prism of Web Intelligence. Communications of the ACM 50(3), 89–94
   (2007)
8. Chapman, G., Marc, R.: The National Information Infrastructure: A Public Interest Opportunity. Computer Professionals for Social Responsibility 11(2), 13–15
   (1993)
9. Collins, F.S., Patrinos, A., Jordan, E., et al.: New Goals for the U.S. Human
   Genome Project: 1998-2003. Science 282(5389), 682–689 (1998)
10. Freeston, M.: The Alexandria Digital Library and the Alexandria Digital Earth
    Prototype. In: Proceedings of the 4th ACM/IEEE-CS joint Conference on Digital
    Libraries (2004)
11. Fayyad, U.M., Piatetsky-Shapiro, G., Smyth, P.: From Data Mining to Knowledge
    Discovery: an overview. In: Advances in Knowledge Discovery and Data Mining
    (1996)

# Reading What Machines *"Think"*

Fabio Massimo Zanzotto and Danilo Croce

University of Rome Tor Vergata
00133 Roma, Italy
{zanzotto,croce}@info.uniroma2.it

**Abstract.** In this paper, we want to farther advance the parallelism between models of the brain and computing machines. We want to apply the same idea underlying neuroimaging techniques to electronic computers. Applying this parallelism, we can address these two questions: (1) how far we can go with neuroimaging in understanding human mind? (*foundational perspective*); (2) can we understand what computers *"think"*? (*applicative perspective*). Our experiments demonstrate that it is possible to believe that both questions have positive answers.

## 1 Introduction

Studies of machines and of living organisms are strongly related. Biological principles have been often used to design machines. In cybernetics [1], artificial adaptive machines, i.e., machines that can control their states, have been studied with respect to the adaptivity of natural living organisms. This relation between studies of living organisms and design of machines is even more strict when we observe computing machines. It is not hard to imagine that the Von Neumann architecture [2] and the neural-based computing architecture originally introduced by Turing [3] have been inspired by concepts coming from the study of the mind and of the brain. Viceversa, also theories of computer architectures inspired some studies of the mind and the brain. Cognitive Psychology [4] uses computing machines as a metaphor for defining models of the human mind. Cognitive Science (see [5]) is even more radical as computing machines are exploited to test and validate models of the mind.

The strict relation between models of the mind and computing machines is fascinating. Staying in this tradition, in this paper, we want to farther advance this parallelism.

*Brains* are often studied using neuroimaging techniques to discover areas related to particular cognitive processes. Neuroimaging techniques are also used to induce activation patterns for high-level cognitive processes related to specific semantic categories [6]. These activation patterns can be used to determine what cognitive process a brain is performing.

*Computing machines* nowadays are extremely complex. These machines perform complex tasks. These tasks seem to be *"cognitive processes"*[1]. For example, manipulating symbols can be seen as a *"cognitive process"*.

---

[1] We often use terms in wrong contexts, e.g., *"cognitive processes"* as related to machines. Yet, we need to make use of human centered terms for describing machines. Whenever we misuse a term, we will indicate it with different characters as we did for *"think"* in the title and for *"cognitive processes"*.

N. Zhong et al. (Eds.): BI 2009, LNAI 5819, pp. 159–170, 2009.

We have then a wonderful opportunity. We can think to observe electronic computers with techniques similar to brainimaging. We have good reasons for doing this as we can address these two questions: (1) how far we can go with neuroimaging in understanding human mind? (*foundational perspective*); (2) can we understand what computers "*think*"? (*applicative perspective*).

This is a fascinating research program and in this introduction we gave only a taste of it. We will better describe our vision in Sec. 2. We describe the parallelism between brains and machines that gives the possibility of applying "*brainimaging*" techniques for electronic computers. We better introduce the two reasons motivating this novel view: a *foundational perspective* and an *applicative perspective*. The rest of the paper is organized as follows. In Sec. 3, we report on the studies that are the background for this research. In Sec. 4, we describe the approximated model for studying the parallelism between brains and machines. In Sec. 5, we report on the experiments for testing the approximated model. We here define a test set that can be used for further experiments. Finally, in Sec. 6, we draw some conclusions on this experience and we plan the future work.

## 2   The Vision

Electronic computers nowadays are extremely complex information processing systems. In some sense, these machines are performing "*cognitive processes*". As previously happened in cognitive science and in cognitive psychology studies, we can imagine the parallelism between computers and minds in the field of *neuroimaging*. Computers as well as brains are the physical objects performing "*cognitive processes*". We hereafter call them "*cognitive physical objects*". As shown in Fig. 1, cognitive tasks activate the "*cognitive physical objects*". In both cases, it is possible to observe the activation of these cognitive objects by taking activation images. We can take these activation images by observing different physical phenomena, e.g., electric or magnetic. The parallelism is now complete. On the *brain* side, we have the brain as the observed *cognitive physical object*, a real cognitive task, and classical brainimaging techniques, i.e., fMRI, as the way of observing the brain activation. On the *electronic computer* side, we have the computer as the *observed cognitive physical object*, a program as the *cognitive task*, and images of the electrical activation of micro-chips as a way of observing the computer activation.

The parallelism we made between brains and computers in the field of neuroimaging opens two possible very interesting research perspectives:

- *foundational perspective*: how far we can go with neuroimaging in understanding human mind?
- *applicative perspective*: can we understand what computers "*think*"?

Both research questions are extremely fascinating.

The *foundational perspective* is extremely important. The aim of some studies in neuroimaging [6,7] is to determine the correlation between high-level cognitive processes and neuroimages. The idea is that processing different conceptual

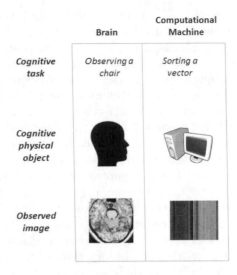

**Fig. 1.** Observing brains and machines

knowledge produces different brain images representing different activation patterns. For example, different conceptual information such as faces, chairs, and houses produces different activation patterns [6]. These correlations can be used for better understanding the way the human brain organizes conceptual knowledge. Yet, an extremely important question is: *how far we can go with neuroimaging in understanding human mind using these methods?*. With the parallelism between brains and machines we have a wonderful opportunity to answer to this question. On the *brain side* (Fig. 1), we have two known variables, i.e., the required cognitive activity and the observed activation pattern, and one unknown variable, i.e., the way the brain is performing the cognitive process. In brain imaging, the aim is to understand and to model the unknown variable. On the *electronic computer* side, there are no unknown variables: the three elements are completely known. This gives a very relevant test-bed. We know exactly how *"knowledge"* is processed in computers and we know exactly the *"cognitive process"* we ask machines to do. If we succeed in studying the correlation between the cognitive process and the activation image in the electronic computer side, we can be confident that the same method can be used on the brain side. We can also answer two additional questions. On the electronic computer side, we can study if better activation image interpretation models produce better correlations between activation images and cognitive activities. For example, what is the effect of knowing that processes are stored as code and data? Does it help in determining the correlation between activation images and the process *"cognitive"* activity? Answering these kinds of questions on the *electronic computer* side can help in determining if clearer separations between brain images related to different cognitive activities correlate with better understanding of the brain cognitive processes.

The *applicative perspective* is also an extremely interesting and unexplored area of research. Using the ideas developed on the *brain* side of the parallelism (Fig. 1), we can try to apply them to the *electronic computer* side. Can we develop technologies that *"read the computer mind"*. This predictive model can have a wide variety of applications, e.g., detecting malicious software, detecting the intentions of hostile computers by looking at their activation patterns. We need specific devices that can capture activation images of computers. We can then study the application of machine learning to induce models that can predict what a computer is doing by analyzing its activation patterns.

The complete realization of the *electronic computer* side of the vision is a long term goal. It requires physical devices to capture the activation state of electronic computers. Yet, as electronic computers are easily and directly observable, it is possible to set up a scenario where we can test the idea. This scenario can help in preparing the ground of the complete research program. This first phase of analysis is the *virtual observation of electronic computers*. We exploit the fact that we can directly observe the memory state of machines and, then, we can draw their activation state. As the observation of the activation state is done through a software system instead of a physical device, we call it *virtual observation*. We will describe this scenario in Sec. 4.

## 3    Background

Categorization is the cognitive ability of classifying objects into concepts. This ability is extremely important for this study as, on both the *brain* side and the *electronic computer* side, we want to study the correlation between activation images (i.e., objects) and cognitive processes (i.e., concepts). The final aim is to develop models that determine the performed cognitive process by observing an activation image. For example, we want to have a model that determines that the brain in Fig. 1 is performing the act of looking at a chair. This should be done only by observing the brain image.

One of the objectives of machine learning is to define models and algorithms that can learn categorization functions from existing training data. Observing some brain images grouped into classes, i.e., grouped according to the cognitive process, machine learning algorithms induce classifiers that can predict the class for a new and unseen brain image. A classification function $C$ is defined as:

$$C : I \to T \tag{1}$$

where $I$ is an instance space and $T$ is the set of possible categories. This classification function will observe objects $i \in I$ assigning a class $t \in T$. The categorization is possible if some regularities appear in the space of the instances $I$. To discover these regularities, we need to observe instances using some feature. These instances are then represented as points in feature spaces $F_1 \times ... \times F_n$ where each $F_i$ is an observable feature. We can then define a function $F$ that maps instances $i$ in $I$ to points in the feature space, i.e.

$$F(i) = (f_1, ..., f_n) \tag{2}$$

This model is generally called feature-value vector and underlies many algorithms.

The field of machine learning has delivered a wide range of algorithms to analyze huge amounts of data in order to find regularities. Supervised (e.g., [8,9]), semi-supervised (e.g., [10]), and weakly supervised (e.g., [11]) algorithms and models are available to automatically learn classifiers or decision making systems. These models are widely and successfully applied in many important fields, e.g., homeland security (e.g. [12]), data mining for business intelligence (e.g. [13]), and computational linguistics [14].

Machine learning algorithms have been used to discover regularities in images of brains performing cognitive and semantic tasks. The work in [7] follows the idea that it is possible to discover regularities in brain images of individuals observing or thinking of objects in the same semantic class such as chairs, houses, etc. (e.g., [6]). Machine learning has been applied to induce brain activation patterns for words where the activation image is not observed. Words with similar meaning should have similar activation patterns. Using corpus linguistics, word similarity is determined comparing their *distributional meaning*, i.e., their vectors of co-occurring words. This is the distributional way of determining the meaning of a word [15]. The induced activation patterns have high predictive performance.

## 4 Virtual Observation of Computational Machines

Electronic computers have a very nice property with respect to our research program. The activity of these machines is observable using software programs. Then, we can simulate the *electronic computer side* of our vision without actually having a physical device to observe the activation state of machines. We can write a software program that snapshots the memory of the machine. These snapshots can then be used to produce activation images as if these were taken from an external device.

Using these virtual observations of the activation states, we can test the overall process of the *electronic computer side*. Then, we can study if it is possible to derive a correlation between the images of the activation states and the performed "*cognitive processes*". For this purpose, we will extract features from activation images to feed machine learning algorithms. Given a set of training examples, i.e., training activation states, associated with different types of "*cognitive activities*", the machine learning algorithm can extract prototypical models of activation for these types of cognitive activities. These latter models can be used to classify novel activation states, i.e., recognize the type of cognitive process that the activation state suggests. If classifiers have good performances with respect to a set of testing activation states, we can conclude that the task of reading "*machines' thoughts*" is reachable using the proposed features. Finally, we can repeat the process using smoothed images of the activation states. Smoothed images better approximate the images produced with physical observation devices. Then, we can determine if the final vision is viable.

In the rest of this section, we first describe the way of capturing the activation state of machines using software programs (Sec. 4.1). Then, we describe the standard features for image classification we used (Sec. 4.2).

## 4.1  Capturing the Activation State

In an electronic computer, we can simulate the capturing of the activation state by directly observing the status of the memory. The way we are doing this then is simple. The aim of this phase is to produce an image representing the activation state of a machine performing a particular "*cognitive task*", e.g., sorting a vector or comparing two strings. We exploit the fact that processes perform "*cognitive activities*". We can define here a "*cognitive activity*" as the execution of a program over input data. Processes are completely represented in memory, i.e., both programs and data are stored in memory. Then, we can directly take snapshots of the memory associated with target processes. These snapshots can be used to build images.

Given a cognitive activity, the procedure for extracting images from this activity is then the following:

- running the process representing the cognitive activity, i.e., the program and the related input data
- stopping the process at given states or at given time intervals $\tau$
- dumping the memory associated with the process
- given a fixed height image and the memory dump, read incrementally bytes of the memory dump and fill the associated RGB pixel with the read values
- eventually, produce a smoothed image

This simple procedure can produce more images for each process related to a cognitive activity.

As it is important to explain which part we are using, here we briefly describe the organization of the process memory (see Fig. 2). The process memory contains: the process control block, the stack, the heap, the data, and the program text. The process control block (PCB) contains the information about its status, i.e., the process identification number, the program counter, the registers, the list of opened files, etc. The stack contains information regarding function calls, passed arguments, and local variables. The heap contains dynamically allocated data, e.g., vectors with a length decided at run-time. The data area contains the statically allocated data, e.g., vectors with fixed length decided at compilation time. Finally, the text area contains the compiled program.

The relative size of the different memory areas is extremely important. The PCB is extremely small with respect to the other areas. The fact that it is not directly observable from an external process is then not relevant. The information loss can be ignored. The other four areas are instead observable. Yet, in general, the data areas (the heap and the data area itself) are bigger than the code area. Thus, a large part of the memory image represents the data areas.

The process memory is then transformed into an image using the following procedure. Let $M(p)$ be the memory dump of the process $p$. The memory dump

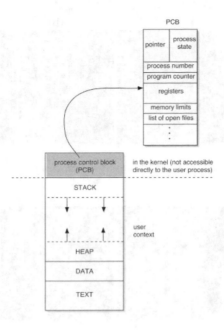

**Fig. 2.** Organization of the memory for a process

is a sequence of bytes, i.e., $M(p) = [b_0, ..., b_m]$. Using this sequence of bytes we can produce an image $I(p)$ in Red-Green-Blue (RGB) coding. The image $I(p)$ is a bi-dimensional array of pixels $p_{i,j}$. Each pixel is three contiguous bytes of the memory image. Given the height $h$ of the image, each RGB pixel has the following RGB values:

$$p_{i,j} = [b_{3(i+h \cdot j)} \ b_{3(i+h \cdot j)+1} \ b_{3(i+h \cdot j)+2}] \tag{3}$$

where the first byte $b_{3(i+h \cdot j)}$ is used for the red component, the second $b_{3(i+h \cdot j)+1}$ for the green one, and the third $b_{3(i+h \cdot j)+2}$ for the blue one. Figure 3 provides an example of two memory images for the "*cognitive task*" of sorting a vector. Figure 3(a) is the process at the initial state and Figure 3(b) is the process that accomplished the task. The smaller stripe on the left of each figure is the code area. This is stable during the process execution. The bigger stripe on the right is the vector. At the beginning of the process (Fig. 3(a)), this area is homogeneous as it shows a random vector. At the end of the process (see Fig. 3(b)), we can see a figure suggesting we have a sorted vector.

Finally, to approximate the physical extraction of the activation images, we use a distortion process for the images. This distortion process allows seeing images where contiguous pixels are merged. This approximates the condition of images captured by a physical scanning device. We cannot expect images of the resolutions given by equation (3). This distortion is called *smoothing* or *blurring*. We use the simplest smoothing model, i.e., the rectangular uniform filter. According to this filter, each pixel in the smoothed image $s_{i,j}$ is a weighted

(a)                           (b)

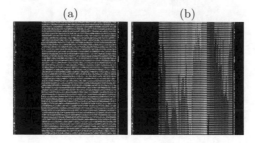

**Fig. 3.** Sorting Process: two activation states

(a)                           (b)

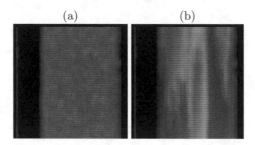

**Fig. 4.** Sorting Process: two distorted activation states

sum of the rectangle $n \times m$ of pixels around the target pixel in the original image. If we use $K = n = m$, $s_{i,j}$ are the smoothed image pixels, and $p_{i,j}$ are the original image pixels, the smoothed pixels are defined as:

$$s_{i,j} = \frac{1}{K^2} \sum_{u=0}^{K-1} \sum_{v=0}^{K-1} p_{i+u-\lfloor \frac{K}{2} \rfloor, j+v-\lfloor \frac{K}{2} \rfloor} \tag{4}$$

This kind of distortion is very interesting as it mixes information extracted from contiguous pixels. For example, the distortion of the images in Fig. 3 is reported in Fig. 4. These latter are obtained using a parameter $K = 10$. This smoothing is particularly relevant as it models what can happen in the situation we have in the final setting, i.e., we are using an external capturing device to extract activation images of electronic computers.

### 4.2    Feature Space for Images

Once we have the complete or the smoothed activation images, we can model them in the selected feature space to finally use the machine learning algorithm to induce the classifiers. We describe here the features we used. The basic idea is to use what is already available for image processing to estimate how far we can go.

We used three major classes of features: chromatic, textures (OP - OGD) and transformation features (OGD), as described in [16]. Chromatic features express the color properties of the image. They determine, in particular, an n-dimensional vector representation of the 2D chromaticity histograms. Texture features emphasize the background properties and their composition. Texture feature extraction, in LTI-Lib [17], uses the steerability property of the oriented gaussian derivatives (OGD) to also generate rotation invariant feature vectors. Transformations are thus modeled by the OGD features. A more detailed discussion of the theoretical and methodological aspects behind each feature set are presented in [16].

# 5   Experimental Evaluation

We have now the possibility of investigating whether or not we can solve the problem of determining the *"cognitive process"* given the activation image. This can help in finding initial answers to the questions issued in Sec. 2 related to the *foundational perspective* and the *applicative perspective*.

The rest of the section is organized as follows. First, we describe the experimental settings (Sec. 5.1). Second, we give the results of the experiments (Sec. 5.2). Finally, we discuss these results (Sec. 5.3).

## 5.1   Experimental Setting

For our experiments, we selected 3 different *"cognitive tasks"*, i.e., 3 types of algorithms: sorting, comparing two strings, and visiting a binary tree. We used *quicksort* as the sorting algorithm and we selected the *levenshtein distance algorithm* for comparing two strings. We implemented these algorithms in 3 different programming languages (c, java, and php) on a linux platform. We have then 3 different *"cognitive tasks"*.

For each of the 3 algorithms in the 3 programming languages, we randomly generated 20 different input data according to the type of algorithm. For the quicksort algorithm, we generated 20 unsorted vectors of $10^5$ elements. For the levenshtein distance algorithm, we generated randomly 20 pairs of strings of 100 characters. For the tree visiting algorithm, we generated 20 trees with $10^4$ nodes and a random node for each tree. We then have 180 different algorithm-data pairs. For each of the algorithm-data pair, we took 3 snapshots: one at the beginning of the execution, one in the middle, and one at the end. We have then 540 instances. We randomly selected 50% of the instances as training and 50% of the instances for testing. We kept the same distribution of the algorithms and of the languages for both training and testing.

We have then defined two classification tasks:

- *per-algorithm task* (*algo*), the final classes are the 3 different algorithms (i.e., sorting, comparing, and visiting), regardless of the programming language of the implementation
- *per-language task* (*lang*), the final classes are the 3 different programming languages (i.e., c, java, and php), regardless of the implemented algorithm

**Table 1.** Classification accuracy over the different experimental settings

| Task | Smoothing | DecTree | NaiveBayes |
|------|-----------|---------|------------|
| algo | no | 80.37 | 64.44 |
| lang | no | 98.89 | 99.25 |
| algo | yes | 81.48 | 62.96 |
| lang | yes | 98.89 | 99.62 |

The *per-algorithm task* is our final task, i.e., understanding whether or not it is possible to determine the *"cognitive task"* performed by the machine. The *per-language task* is instead a control test. We want to see if it is possible to determine the *"cognitive substrate"* where the *"cognitive task"* is performed. As this task seems to be easier and it is similar to the *per-algorithm task*, we want also to see if it is solvable with the feature space for images we are using.

For the two classification tasks, we prepared two different settings according to the smoothing applied to the final images:

- *no-smoothing*, images are kept with the highest quality available;
- *smoothed*, images are smoothed according to the equation (4).

We need these two settings to determine if the performance of the two classification tasks is affected by a degradation of the images. As we already discussed, the degradation approximates the real operational conditions where the captured image cannot have the quality of the single byte of memory.

Finally, as machine learning algorithms we used a decision tree learner [8] (*DecTree*) and a probabilistic model, i.e., the naive bayes (*NaiveBayes*). We selected these two types of algorithms because they behave completely differently. The decision tree learning algorithm recursively selects features. At each step, the one selected is the best discriminatory one. The pruning done in the J48 algorithm also performs a feature selection. Yet, the naive bayes algorithm uses and *weights* all the features in the probabilistic model. These two algorithms then give the possibility of analyzing performances in two completely different setting. We used the implementation given in [18].

## 5.2    Results

The results of the experiments are reported in Tab. 1. The first column describes the tasks that have been analyzed. The second column reports if the smoothing has been applied. The third column reports the accuracy obtained using the decision tree learning algorithm. Finally, the third column reports the accuracy obtained using a naive bayes probabilistic learning algorithm.

## 5.3    Discussion

As expected, the task of deciding the language of the process is simple and solvable. Accuracies are extremely high. There is no difference in performance either between the machine learning algorithms or between the quality of the images. Programming languages produce different organizations of the memory of the processes.

Deciding the algorithm performed by a process is instead more complex. Yet, results are encouraging. The performance of the decision tree learning algorithm is above 80%. The performance is even higher for smoothed images. The differences between the naive bayes algorithm and the decision tree learning suggests that some features are more informative than others. These features are even more important for smoothed images. This increase in performance is unexpected. The behavior of the naive bayes algorithm is instead more predictable as classifiers learnt and applied on plain images performed better than those learnt and applied on smoothed images.

These are initial answers to the two questions issued in Sec. 2. As discussed in the next section, we still need to investigate more complex datasets to generilze these initial findings.

## 6    Conclusions and Future Work

We introduced a new vision that can both help in answering questions in neuroimaging and produce novel applications in the computer field. The results of the experimental evaluation suggest we can read what computers *"think"* as we can positively predict *"cognitive processes"* from activation images. Then, we can have positive expectations of the foundational perspective and the application perspective. Yet, the research program we started is still at the beginning and many questions still have to be addressed. For this study we investigated simple *"cognitive processes"*. We need to scale-up to different datasets, e.g., datasets containing activation images of word processors and image processors. Then, we have to attack the problem of determining which processes are active in a given memory. Finally, we have to figure out physical devices to directly capture activation images from the electronic computer.

## Acknowledgement

We would like to thank Marco Cesati for his precious advices on the organization of the memory in the Linux kernel.

## References

1. Wiener, N.: Cybernetics: Or the Control and Communication in the Animal and the Machine. MIT Press, Cambridge (1948)
2. von Neumann, J.: First draft of a report on the edvac. IEEE Ann. Hist. Comput. 15(4), 27–75 (1993)
3. Turing, A.: Intelligent Machinery. In: Meltzer, B., Michie, D. (eds.) Machine Intelligence, vol. 5, pp. 3–23. Edinburgh University Press, Edinburgh (1969)
4. Neisser, U.: Cognitive Psychology. Appleton-Century-Crofts, New York (1967)

5. Von Eckardt, B.: What is cognitive science? MIT Press, Cambridge (1993)
6. Ishai, A., Ungerleider, L.G., Martin, A., Schouten, J.L., Haxby, J.V.: Distributed representation of objects in the human ventral visual pathway. In: Proc. Natl. Acad. Sci. U S A, vol. 96(16), pp. 9379–9384 (1999)
7. Mitchell, T.M., Shinkareva, S.V., Carlson, A., Chang, K.M., Malave, V.L., Mason, R.A., Just, M.A.: Predicting human brain activity associated with the meanings of nouns. Science 320(5880), 1191–1195 (2008)
8. Quinlan, J.: C4:5:programs for Machine Learning. Morgan Kaufmann, San Mateo (1993)
9. Vapnik, V.: The Nature of Statistical Learning Theory. Springer, Heidelberg (1995)
10. Blum, A., Mitchell, T.: Combining labeled and unlabeled data with co-training. In: COLT: Proceedings of the Conference on Computational Learning Theory. Morgan Kaufmann, San Francisco (1998)
11. Dempster, A., Laird, N., Rubin, D.: Maximum likelihood from incomplete data via the em algorithm. Journal of the Royal Statistical Society, Series B 39(1), 1–38 (1977)
12. Goan, T., Mayk, I.: Improving information exchange and coordination amongst homeland security organizations. In: Proceedings of the 10th International Command and Control Research and Technology Symposium, ICCRTS (2005)
13. Bose, I., Mahapatra, R.K.: Business data mining: a machine learning perspective. Information & management 39, 211–225 (2001)
14. Zanzotto, F.M., Moschitti, A.: Automatic learning of textual entailments with cross-pair similarities. In: Proceedings of the 21st Coling and 44th ACL, Sydney, Australia, pp. 401–408 (July 2006)
15. Harris, Z.: Distributional structure. In: Katz, J.J., Fodor, J.A. (eds.) The Philosophy of Linguistics. Oxford University Press, New York (1964)
16. Alvarado, P., Doerfler, P., Wickel, J.: Axon$^2$ - a visual object recognition system for non-rigid objects. In: IASTED International Conference-Signal Processing, Pattern Recognition and Applications (SPPRA), Rhodes, IASTED, pp. 235–240 (July 2001)
17. Alvarado, P., Doerfler, P.: LTI-Lib - A C++ Open Source Computer Vision Library. In: Kraiss, K.F. (ed.) Advanced Man-Machine Interaction. Fundamentals and Implementation, pp. 399–421. Springer, Dordrecht (2006)
18. Witten, I.H., Frank, E.: Data Mining: Practical Machine Learning Tools and Techniques with Java Implementations. Morgan Kaufmann, Chicago (1999)

# Using SVM to Predict High-Level Cognition from fMRI Data: A Case Study of 4*4 Sudoku Solving

Jie Xiang[1,2], Junjie Chen[1], Haiyan Zhou[2], Yulin Qin[2,3],
Kuncheng Li[4], and Ning Zhong[2,5]

[1] College of Computer and Software, Taiyuan University of Technology, China
[2] The International WIC Institute, Beijing University of Technology, China
[3] Dept of Psychology, Carnegie Mellon University, USA
[4] Dept of Radiology, Xuanwu Hospital Capital University of Medical Sciences, China
[5] Dept of Life Science and Informatics, Maebashi Institute of Technology, Japan
yulinqin@gmail.com, zhong@maebashi-it.ac.jp

**Abstract.** In this study, we explore the approach using Support Vector Machines (SVM) to predict the high-level cognitive states based on fMRI data. On the base of taking voxels in the brain regions related to problem solving as the features, we compare two feature extraction methods, one is based on the cumulative changes of blood oxygen level dependent (BOLD) signal, and the other is based on the values at each time point in the BOLD signal time course of each trial. We collected the fMRI data while participants were performing a simplified 4*4 Sudoku problems, and predicted the complexity (easy vs. complex) or the steps (1-step vs. 2-steps) of the problem from fMRI data using these two feature extraction methods, respectively. Both methods can produce quite high accuracy, and the performance of the latter method is better than the former. The results indicate that SVM can be used to predict high-level cognitive states from fMRI data. Moreover, the feature extraction based on serial signal change of BOLD effect can predict cognitive states better because it can use abundant and typical information kept in BOLD effect data.

## 1   Introduction

Methods in machine learning, such as classification, have been introduced into fMRI (functional Magnetic Resonance Imaging) data mining to investigate neural cognitive mechanism and decode mind states. In early years, Haxby and colleagues used Support Vector Machines (SVM) to train classifiers to predict whether a participant was watching a shoe or a bottle [1], and a human face or an object [2], which provided more information about semantic representation in human brain. Since then, more and more researchers have investigated the methods of machine learning to analyze fMRI data collected in visual or auditory perception tasks. Kamitani and Tong's work showed that different angles of gratings in vision could be distinguished through fMRI data [3]. By analyzing

N. Zhong et al. (Eds.): BI 2009, LNAI 5819, pp. 171–181, 2009.
© Springer-Verlag Berlin Heidelberg 2009

the fMRI data with LS-SVM (Least Square Support Vector Machine) algorithm, Formisano et al. could predict "who" (male or female) was saying "what" (which German vowels) through fMRI classification [4]. Haynes et al. even used fMRI data from visual cortex to classify and predict human's quick stream of consciousness [5]. In a serial of studies, Mitchell and colleagues applied the methods of fMRI classification in several cognitive tasks: (1) reading a set of words belonging to six types of semantics (such as tools, fruits, and so on), (2) reading two types of sentences (explicit vs. ambiguous sentences), (3) viewing sentences and pictures [6,7,8,9]. As a type of data-driven method, classification emphasizes the mapping between observed fMRI data and the cognitive states. Although many researchers used this machine learning method to explore the states of human cognition, most of them focused on perception, but not on high-level cognition with complex information processing, such as problem solving. Different problem tasks may be with similar visual stimulus but need very different ways to solve them. If machine learning approach can be used to predict high-level cognition, this approach might shed light on how the brain processes information, and on mind reading needed in brain-machine interface development. In this study, we explore the approaches from fMRI data to identify the type of the problem the participant was solving.

Several methods of classification have been used in previous studies. Multi-voxel pattern analysis (MVPA) used in Haxby's study addressed the issue about classifier selection [1, 2]. In Kendrick's research, a Gabor Wavelet Pyramid model was adopted to predict the pictures when participants were watching a large number of nature images [10]. Hasson used the information of intersubject synchronization to judge whether a participant was watching human faces or buildings in a film clip [11]. Sato and colleagues compared the methods of SVM and MLDA's (Maximum Uncertainty Linear Discrimination Analysis) performances of predicting human cognitive states in hearing, vision, finger exercises and other tasks [12]. They indicated that all brain areas were required for MLDA, but much less brain areas for SVM. In addition, with the less areas involved in, the classification performances were higher in SVM than that in MLDA. So it seems that the method of SVM had the superiority in classification. We have also tried some different classifiers (including back propagation of neural network and Gaussian Naive Bayes) and also find that SVM is better than others in our study. We will report how we implemented SVM with the emphasis on feature extraction in later sections.

## 2    Method

### 2.1    Task and Materials

Event related fMRI data were recorded while participants were solving simplified 4*4 sudoku tasks. Sudoku is a combinatorial number-placement puzzle, the goal of the puzzle is to fill a $4 \times 4$ grid so that each column, each row, and each of the four $2 \times 2$ boxes contains the digits from 1 to 4 only one time each. As shown in Figure 1, in this study, we simplify the puzzle and ask participants to give the answer of the cell marked with '?'. It was a 2 × 2 designed experiment with two 2-level factors: steps (1-step vs. 2-steps) and complexity (simple vs. complex). There are totally four types of tasks, in a task of 1-step, participants only needed to find the answer of the cell with a mark '?' (e.g., Figure 1(a) and 1(b)); while in a task of 2-steps, participants had to find the answer of the cell with mark '*' before they could find the answer of the cell with mark '?' (e.g., Figure 1(c) and 1(d)). On the other hand, in a simple task, the participants only needed to check one column, one row, or one box to find the answer for '?' (see Figure 1(a) for an example of only checking one box needed); but in a complex task, the participants had to check column and row and/or box (see Figure 1(b) for an example). These tasks had little differences in vision stimuli, but had great differences in the problem solving processes, such as, problem representation and memory retrieval for heuristics.

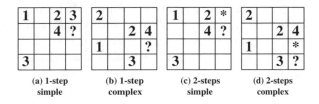

(a) 1-step   (b) 1-step   (c) 2-steps   (d) 2-steps
simple       complex       simple         complex

**Fig. 1.** Examples of materials

As shown in Figure 2, a trial of the experiment starts with a red star shown for 2 seconds as warning (the stimulus is visually shown on a black screen), the participants solve the puzzle in the following period of 20 seconds in maximum. When participants find the answer of '?', they are asked to press a button immediately and speak out the answer in a 2 seconds period. Participants are encouraged to finish the problem as correctly and quickly as possible. After that, the correct answer is provided in the screen for 2 seconds as a feedback. Then there is a 10 seconds of inter-trial interval (ITI; a white cross shown on the screen) and the participants are asked to take rest in this period.

There were 5 sessions each with 48 or more trials and each session involved 4 types of tasks randomly selected with equal probability.

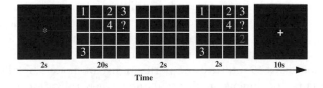

**Fig. 2.** The protocol of a scan trial

## 2.2   Scan Protocols and Data Preprocessing

The images were acquired on a 3.0T MR scanner (Siemens Trio+timGenmeny) NMR equipment and a SS-EPI (single shot echo planar imaging) sequence sensitive to BOLD (blood oxygen level dependent) signals was used to acquire the fMRI data. The functional images were acquired with the following parameters: TR = 2000 ms, TE = 30 ms, flip angle = 90$^o$, FOV = 200mm × 200mm, matrix size = 64 × 64, slice thickness = 3.2 mm, slice gap = 0 mm, and 32 axial slices with AC-PC on the 10th slice from the bottom of brain.

19 college students (9 males and 10 females) from Beijing University of Technology were scanned after obtaining informed consent. The average age of participants was 22.8, and all of them were right-handed native Chinese speakers.

Data preprocessing (e.g., motion correction) and statistical analysis were performed with NeuroImaging Software package (NIS, kraepelin.wpic.pitt.edu/nis/). 3 participants were excluded due to head movement exceeding 5mm. All images were coregistered to a common reference structural MRI image and smoothed with a 6 mm full-width half-maximum three-dimensional Gaussian filter. For each voxel and each trial, the percent signal change relative to the baseline (two scans before the stimulus onset scan of the trial) was calculated and used as fMRI data in classification.

## 2.3   Classification Based on SVM

In general, the process of classification is to find a function mapping the fMRI data to a specific cognitive state through training datasets, and then to use this mapping to determine the unknown cognitive state from fMRI data. The function can be formalized as follows:

$f$ : fMRI-sequence (t, t+n) → $Y$

where fMRI-sequence (t, t+n) is the observed fMRI data during the interval from time t to t+n, $Y$ is a finite set of cognitive states to be discriminated. In this study, we performe two types of classifications, one is step classification (1-step tasks vs. 2-steps tasks), and the other is complexity classification (simple tasks vs. complex tasks).

## Feature (Voxel) Selection

Feature selection is the technique, commonly used in pattern recognition, of selecting a subset of relevant features for building learning models. The performance of pattern classification typically depends on feature selection in which a set of features is selected that contains enough information to perform the classification. By removing most irrelevant and redundant features from the data, feature selection helps to improve the performance of models by alleviating the effect of the curse of dimensionality, enhancing generalization capability and speeding up learning process. In the fMRI data classification, feature selection is equivalent to voxel selection. Typically, voxels can be selected based on anatomical or functional knowledge.

In the present study, voxels in several predefined regions are selected as features according to ACT-R (Adaptive Control of Thought-Rational) theory [15, 16]. As a cognitive architecture, ACT-R assumes that cognition emerges through the interaction of a set of modules, 8 brain regions are mapped to these specific modules. In this study, voxels in 5 bilateral brain regions that are related to the sudoku solving are selected to build our model. The 5 bilateral brain regions are as followed: lateral inferior prefrontal cortex (PFC) , centered at Talairach coordinates x = ±40, y = 21, z = 21, reflect retrieval of information in declarative module; posterior parietal cortex (PPC) , centered at x = ±23, y = -64, z = 34, reflect changes to problem representations in imaginal module; anterior cingulate cortex (ACC), centered at x = ±5, y = 10, z = 38, control various stages of processing and prevent the problem solving state from distracting from the goal; caudate, centered at x = ±15, y = 9, z = 2, play an action-selection role; and fusiform gyrus (FG), centered at x = ±42, y = -60, z = -8, engage in visual processing [14, 15, 16]. 748 voxels in these predefined regions were selected including bilateral PFC (5 × 5 × 4, i.e., 5 voxels wide, 5 voxels long, and 4 voxels high), bilateral PPC region (5 × 5 × 4), bilateral FG region (5 × 5 × 3), bilateral ACC region (3 × 5 × 4), bilateral caudate region (4 × 4 × 4).

## Feature (Value) Extraction

How to extract feature values to build classifier's input vectors is also a key factor to affect the performance of classification. According to hemodynamic response function, the BOLD signal changes by cognition increases to arrive at the peak values after 4-8 seconds from the cognitive stimuli, then decreases gradually, and the BOLD signal lasts about 12-14 seconds. So how to abstract the characteristic of BOLD signal change is very important. In general, as depicted in Figure 3,

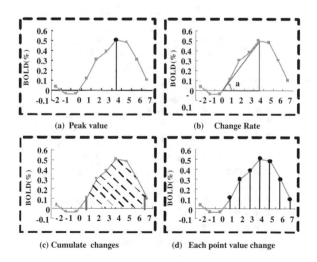

(a)  Peak value    (b)  Change Rate

(c) Cumulate changes    (d)  Each point value change

**Fig. 3.** Features used to describe BOLD effect

different features are used to describe the pattern of BOLD effect, such as peak time, peak value, change rate, cumulative changes, and bold change value at each time point. Among these features, cumulative changes might describe the change rate, peak value, and peak time synthetically.

In this paper, we compare two feature extracting methods for encoding fMRI-sequence (t, t+n) as inputs to the classifier. The first one is to extract cumulative BOLD changes of each voxel located in predefined regions, in which the input vector is a 748-dimensional vector. Supposing $\mathbf{V} = [v_i] = (v_1, v_2, \ldots, v_{748})$, where $v_i$ is the sum of BOLD changes from scan2 to scan7 of the ith voxel. The second one is to extract the change value at each time point of BOLD time course, in which the classifier's input vector is a 6-dimensional vector. Supposing $\mathbf{T} = [t_j] = (t_1, t_2, \ldots, t_6)$, where $t_j$ means the BOLD signal change value at the j+1 scan after stimulus onset ($t_1$ corresponding to scan2, $t_2$ corresponding to scan 3, and so on).

## Classification

Two classifiers based on SVM are compared in the present study, one is called multi-voxel-classifier (MVC), in which the 748-dimensional vector of MVC is used as the future vector of the classifier. The other is called single-voxel based voter classifier (SVVC), in which the 6-dimensional vector for each selected voxel

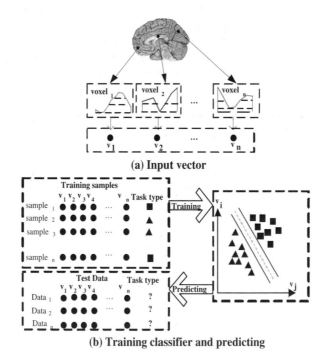

(a) Input vector

(b) Training classifier and predicting

**Fig. 4.** The scheme of MVC

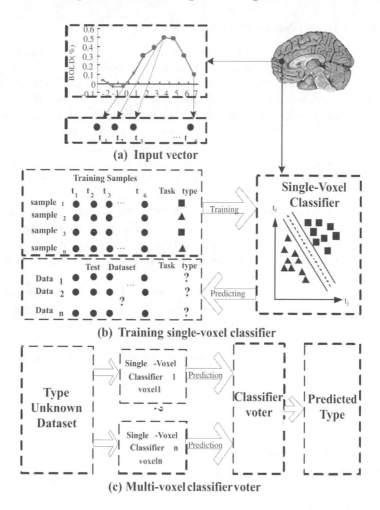

**Fig. 5.** The scheme of SVVC

is the input of each single-voxel classifier. The details of the two kinds of classification are given in the following sub-sections.

## Multi-Voxel-Classifier with the Sum of BOLD Signal Change (MVC)

As shown in Figure 4, the input vectors of the MVC consist of the sum of BOLD changes of voxels in predefined regions, the classifier are trained from the vectors of training samples along with corresponding labels. Once trained, the classifier can be used to predict labels for a test set. The predicted labels are then compared to the true labels and the accuracy of the classifier can be computed.

**Single-Voxel Based Voter Classifier (SVVC)**

As shown in Figure 5, a set of 6-dimensional features vectors along with their labels of a voxel is given to train a single-voxel classifier to form an optimal discrimination hyperplane which means each single-voxel classifier after training can predict the unknown type of test data. Because of a lot of noises in fMRI data, however, the result predicted only by one voxel is unacceptable. This is only the first stage. The second stage treat each single-voxel classifier as a voter, and the final decision is made by calculating the weighted sum of the classification of all single-voxel-classifiers. The weight is corresponding to the performance of classifications of each classifier, the higher classification accuracy a single-voxel-classifier obtained, the higher weight will be assigned to it.

The classifiers mentioned above are all SVM classifiers, and the LIBSVM toolkit [13] based on MATLAB platform was used to train these classifiers. The parameters of training are as follows: svm_type = nu_SVC, kernel_type = RBF, nu = 0.5, C and gamma were adjusted in given interval [-10, 10] to get optimized performance.

The Leave-One-Session-Out method was adopted to evaluate the validity of the classifiers. As there were 5 scan sessions for each participant, we randomly chose 4 sessions data to train the classifier and used the data in the left session to test the classifier to get one validation. There were totally 5 validations, the final accuracy for each participant was the average of the accuracy values in these 5 validations.

# 3   Results

We performed 2 kinds of classification tasks, one was step classification (1-step tasks vs. 2-steps tasks), and the other was complexity classification (simple tasks vs. complex tasks).

## 3.1   Results of MVC

The results of classification based on MVC for all subjects and the average accuracy are shown in Figure 6(a), in which the step classifiers denoted in grey columns, the highest accuracy reaches to 80.9% in individual participants, and the average accuracy across 16 participants is 70.3% [SD = 0.04], which is significantly higher than the random classification chance level of 50%. As to the complexity classifiers denoted in dark columns, the highest accuracy is 68.7% in individual participants, and the average accuracy across 16 participants is 63.3% [SD = 0.03], which is significantly higher than the random classification chance level of 50%. The results show that the classification in step is more effective than the one in complexity.

## 3.2   Results of SVVC

The results of classification based on SVVC for all subjects and the average accuracy are shown in Figure 6(b), in which the step classifiers denoted in grey

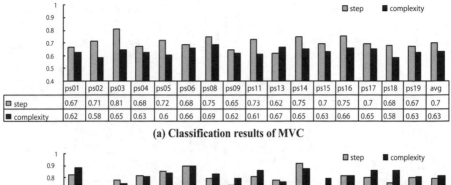

| | ps01 | ps02 | ps03 | ps04 | ps05 | ps06 | ps08 | ps09 | ps11 | ps13 | ps14 | ps15 | ps16 | ps17 | ps18 | ps19 | avg |
|---|---|---|---|---|---|---|---|---|---|---|---|---|---|---|---|---|---|
| □ step | 0.67 | 0.71 | 0.81 | 0.68 | 0.72 | 0.68 | 0.75 | 0.65 | 0.73 | 0.62 | 0.75 | 0.7 | 0.75 | 0.7 | 0.68 | 0.67 | 0.7 |
| ■ complexity | 0.62 | 0.58 | 0.65 | 0.63 | 0.6 | 0.66 | 0.69 | 0.62 | 0.61 | 0.67 | 0.65 | 0.63 | 0.66 | 0.65 | 0.58 | 0.63 | 0.63 |

**(a) Classification results of MVC**

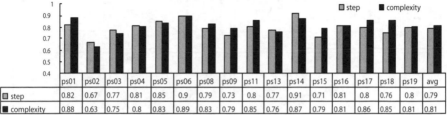

| | ps01 | ps02 | ps03 | ps04 | ps05 | ps06 | ps08 | ps09 | ps11 | ps13 | ps14 | ps15 | ps16 | ps17 | ps18 | ps19 | avg |
|---|---|---|---|---|---|---|---|---|---|---|---|---|---|---|---|---|---|
| □ step | 0.82 | 0.67 | 0.77 | 0.81 | 0.85 | 0.9 | 0.79 | 0.73 | 0.8 | 0.77 | 0.91 | 0.71 | 0.81 | 0.8 | 0.76 | 0.8 | 0.79 |
| ■ complexity | 0.88 | 0.63 | 0.75 | 0.8 | 0.83 | 0.89 | 0.83 | 0.79 | 0.85 | 0.76 | 0.87 | 0.79 | 0.81 | 0.86 | 0.85 | 0.81 | 0.81 |

**(b) Classification results of SVVC**

**Fig. 6.** Classification results

columns, the highest accuracy reaches high to 91.49% in individual participants, and the average accuracy across 16 participants is 79.4% [SD = 0.06]. As to the complexity classifiers, denoted in dark columns, the highest accuracy is 89% in individual participants, and the average accuracy across 16 participants is high to 81.3% [SD = 0.06]. The accuracy of SVVC is higher than MVC significantly $[F(1, 33) = 99.651, p < 0.001]$.

## 4   Discussion

It is well known, classification performance depends on features selection, and suitable features can improve the classifier performance effectively. The high accuracy in our classification demonstrat that selecting PPC, PFC, ACC, FG, caudate as features is suitable for classifying high-level cognitive states. Significant BOLD pattern differences as shown in Figure 7 can explain the high accuracy in our classification. Because these regions are related to solving the 4*4 Sudoku, significant task differences in BOLD signal can improve the accuracy of classification when these regions are selected as features. The high classification accuracy might indicate that these regions take part in problem solving. Hence, the classification method might be used to analyze activated brain regions in cognitive tasks. Researchers have already used this pattern classification in fMRI data analysis to explore brain activity [1, 2]. It is also interesting to see that although the same features are selected in two methods mentioned above, the accuracy of SVVC is higher than MVC significantly. This is due to information lossless by feature extracting from BOLD effect time course in SVVC.

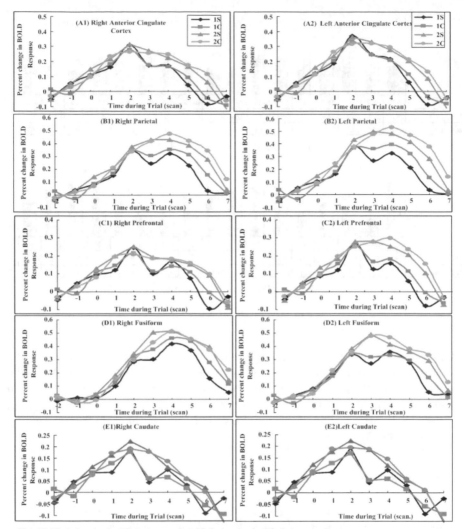

Note (1) 1S: 1-step simple; 1C: 1-step complex; 2S: 2-step simple; 2C: 2-step complex.
(2) 0: onset scan; -1,-2: the first scan and second scan before the stimulus scan onset.

**Fig. 7.** Different BOLD pattern in predefined regions

To conclude, it is possible that using BOLD signal time courses of selected ROIs, including PPC, PFC, ACC, FG, and caudate, to predict the mental states of high-level cognitive processes from fMRI data. To our best knowledge this study is the first one reported in using machine learning methods to decode cognition during problem solving. Until now, this is only a preliminary study, many problems still have to be addressed in future work, such as how to use this kind of method to clarify the roles of brain regions in problem solving, how to understand and explain the performances of classification in depth.

## Acknowledgement

The authors would like to thank Shengfu Lv and Jian Yang for their comments on this paper,the authors also would like to thank Lijuan Wang and Fenfen Wang for their help in experiments. This work was supported by the Natural Science Foundation of China (No. 60875075).

## References

1. Norman, K.A., Polyn, S.M., Detre1, G.J., Haxby, J.V.: Beyond Mind-reading: Multi-voxel Pattern Analysis of fMRI Data. Trends in Cognitive Sciences 10(9), 424–430 (2006)
2. Haxby, J.V., Gobbini, M.I., Furey, M.L., Ishai, A., Schouten, J.L., Pietrini, P.: Distributed and Overlapping Representations of Faces and Objects in Ventral Temporal Cortex. Science 293, 2425–2430 (2001)
3. Kamitani, Y., Tong, F.: Decoding the Subjective Contents of the Human Brain. Nature Neuroscience 8, 679–685 (2005)
4. Formisano, E., Martino, F.D., Bonte, M., Goebel, R.: Who Is Saying What? Brain-Based Decoding of Human Voice and Speech. Science 322, 970–973 (2008)
5. Haynes, J.D., Rees, G.: Predicting the Stream of Consciousness from Activity in Human Visual Cortex. Current Biology 15, 1301–1307 (2005)
6. Mitchell, T.: Learning to Decode Cognitive States from Brain Images. Machine Learning 57, 145–175 (2004)
7. Mitchell, T., Shinkareva, S.V., Carlson, A., Chang, K.M., Malave, V.L., Mason, R.A., Just, M.A.: Predicting Human Brain Activity Associated with the Meanings of Nouns. Science 320, 1191–1195 (2008)
8. Pereira, F.: Beyond Brain Blobs: Machine Learning Classifiers as Instruments for Analyzing Functional Magnetic Resonance Imaging Data, vol. 12. School of Computer Science Carnegie Mellon University, Pittsburgh (2007)
9. Pereira, F., Mitchell, T., Botvinick, M.: Machine Learning Classifiers and fMRI: A tutorial overview. NeuroImage 45, S199–S209 (2009)
10. Kay, K.N., Naselaris, T., Prenger, R.J., Gallant, J.L.: Identifying Natural Images from Human Brain Activity. Nature 452, 352–355 (2008)
11. Hasson, U., Nir, Y., Levy, I., Fuhrmann, G., Malach, R.: Intersubject Synchronization of Cortical Activity During Natural Vision. Science 303, 1634–1640 (2004)
12. Sato, J.R., Fujita, A., Thomaz, C.E., Martin Mda, G., Mour?o-Miranda, J., Brammer, M.J., Junior, E.A.: Evaluating SVM and MLDA in the Extraction of Discriminant Regions for Mental State Prediction. NeuroImage 46, 105–114 (2009)
13. Chang, C.C., Lin, C.J.: LIBSVM: a library for support vector machines (2001), http://www.csie.ntu.edu.tw/cjlin/libsvm
14. Qin, Y., Sohn, M.H., Anderson, J.R., Stenger, V.A., Fissell, K., Goode, A., Carter, C.S.: Predicting the Practice Effects on the Blood Oxygenation Level-Dependent (BOLD) Function of fMRI in a Symbolic Manipulation Task. In: Proceeding of the National Academy of Sciences of the United States of America, vol. 100, pp. 4951–4956 (2003)
15. Anderson, J.R.: How Can the Human Mind Occur in the Physical Universe? Oxford University Press, USA (2007)
16. Anderson, J.R., Bothell, D., Byrne, M.D., et al.: An Integrated Theory of the Mind. Psychological Review 111(4), 1036–1060 (2004)

# Data-Brain Modeling for Systematic Brain Informatics

Jianhui Chen[1] and Ning Zhong[1,2]

[1] International WIC Institute, Beijing University of Technology, Beijing,
100124, P.R. China
[2] Dept of Life Science and Informatics, Maebashi Institute of Technology,
Maebashi-City, 371-0816, Japan
chenjhnh@gmail.com, zhong@maebashi-it.ac.jp

**Abstract.** In order to understand human intelligence in depth and find
the cognitive models needed by Web Intelligence (WI), Brain Informatics (BI) adopts systematic methodology to study human "thinking centric" cognitive functions, and their neural structures and mechanisms
in which the brain operates. For supporting systematic BI study, we
propose a new conceptual brain data model, namely Data-Brain, which
explicitly represents various relationships among multiple human brain
data sources, with respect to all major aspects and capabilities of human information processing systems (HIPS). On one hand, constructing
such a Data-Brain is the requirement of systematic BI study. On the
other hand, BI methodology supports such a Data-Brain construction.
In this paper, we design a multi-dimension framework of Data-Brain and
propose a BI methodology based approach for Data-Brain modeling. By
this approach, we can construct a formal Data-Brain which provides a
long-term, holistic vision to understand the principles and mechanisms
of HIPS.

## 1 Introduction

Brain Informatics (BI) [13,14] is a new interdisciplinary field to study human
information processing mechanism systematically from both macro and micro
points of view by cooperatively using experimental/computational cognitive neuroscience and Web Intelligence (WI) and data mining centric advanced information technologies. It can be regarded as brain sciences in the WI centric IT
age [12,14].

The capabilities of human intelligence can be broadly divided into two main
aspects: perception and thinking. Though it is not appropriate to say that thinking centric cognitive functions are superior to perception oriented functions,
thinking centric functions obviously reflect more intelligence of HIPS than perception oriented functions. Thus, for getting cognitive models needed by WI,
our BI study focuses on the "thinking centric" cognitive functions, which are
complex and involved with multiple inter-related cognitive functions with respect to activated brain areas for a given task, and neurobiological processes of

N. Zhong et al. (Eds.): BI 2009, LNAI 5819, pp. 182–193, 2009.

spatio-temporal features related to the activated areas. Aiming at the characteristics of thinking centric functions, BI emphasizes on a *systematic* approach for investigating human information processing mechanisms, including measuring, collecting, modeling, transforming, managing, and mining multiple human brain data obtained from various cognitive experiments by using fMRI (functional magnetic resonance imaging) and EEG (electroencephalogram).

Such systematic study includes four core issues: systematic investigation of human thinking centric mechanisms, systematic design of cognitive experiments, systematic human brain data management, and systematic human brain data analysis and simulation. For supporting this systematic study, BI attempts to capture new forms of collaborative and interdisciplinary work. New kinds of BI methods and global research communities will emerge, through infrastructures on the Wisdom Web[11] and knowledge grids, that enable high speed and distributed, agent-based large-scale analysis, simulations and computations, and radically new ways of sharing data and knowledge.

A BI portal [12] is just such an infrastructure which implements data storage, sharing, and utilization with a long-term, holistic vision for systematic data management. It is oriented to analysis and simulation, so conceptual modeling of brain data, i.e., Data-Brain modeling [2], is a core issue in BI portal construction. Systematic BI methodology also supports such a Data-Brain construction.

This paper presents a case study on Data-Brain modeling based on BI methodology. The rest of this paper is organized as follows. Section 2 discusses background and related work. Section 3 gives the definition of Data-Brain and its multi-dimension framework. Section 4 describes a BI methodology based approach for Data-Brain modeling. Finally, Section 5 gives concluding remarks.

## 2   Background and Related Work

As a typical "data" science, the key questions of recent Brain Informatics study are how to obtain high quality of experimental data, how to manage such huge multimedia data, as well as how to analyze such data for discovering new brain related knowledge. Effective data management is the base of BI study. Over the last decade, researchers have focused their efforts on constructing various brain databases for supporting brain data management. These brain databases can store microcosmic structure data [21] and macroscopic structure data [16], respectively. As two kinds of important macroscopic brain data, both fMRI and EEG are also focused. The researchers constructed many related brain databases, such as fMRIDC [18], AED [15], all of which are oriented to data storage and sharing.

The construction of traditional databases often adopts a top-down course, from conceptual modeling to physical modeling. But in the existing studies of brain databases, conceptual modeling of brain data is often neglected. Because most of data in the existing brain databases come from separated studies, the relationships among data are lack. Conceptual modeling can only be used to describe the limited relationships among data in the same experiment task, so it

cannot get enough attention. Because of the neglect of conceptual data modeling, it is difficult to support systematic BI study based on these brain databases
since the relationships among data cannot be explicitly represented and applied.
For example, systematic data analysis needs to adopt agent-enriched human
brain data mining for multi-aspect analysis and simulation on multiple human
brain data sources. This needs not only systematic data storage based on the
relationships among data, but also a formal conceptual model which explicitly
describes the relationships among data to guide agent computing. Hence, systematic BI methodology needs the study of conceptual modeling of brain data,
i.e., Data-Brain modeling.

The complexity of human brain leads that BI study needs global cooperation.
Thus, more and more researchers focus their efforts on connecting decentralized
brain databases by network or grid infrastructure to construct various resource
networks/grids for supporting global cooperation. The international neuroin-
formatics network is just such a resource network, which contains many brain
database nodes, such as Brain Bank [17]. At the network/grid level, conceptual
data models have the wider scope of applications, including not only off-time
applications, such as supporting the design of database schemas, but also on-
time applications, such as providing formal knowledge sources. Obviously, the
traditional graphical modeling languages, such as the Entity-Relationship (ER)
model [1], cannot meet all the requirements of new applications.

At present, ontologies are widely applied on network/grid based systems [6].
Both ontologies and data models are partial accounts of conceptualizations [9],
and the common features between them have gotten attention [5]. Though some
researchers focus on differentiating ontologies from conceptual data models [4],
the various applications of ontologies, especially the applications in resource
networks/grids, still make the ontology be a new effective approach for formally
modeling data related domain knowledge at the conceptual level. Hence, ontology can be regarded as a new approach of conceptual data modeling at the
network/grid level.

## 3 Data-Brain

### 3.1 What Is a Data-Brain?

The Data-Brain is a conceptual brain data model, which represents functional relationships among multiple human brain data sources, with respect to all major
aspects and capabilities of HIPS, for systematic investigation and understanding of human intelligence. On one hand, developing such a Data-Brain is a core
research issue of BI. Systematic BI study needs a Data-Brain to describe multi-
aspect data related knowledge for supporting systematic data storage, sharing,
and utilization. Based on this way, it provides a long-term, holistic vision to
uncover the principles and mechanisms of underlying HIPS. On the other hand,
BI methodology supports such a Data-Brain construction. As a network/grid
level of conceptual data model, the Data-Brain can adopt an ontological modeling approach based on BI methodology. In other words, the Data-Brain goes

beyond specificity of a certain application and straightly models related domain knowledge based on multiple core issues of BI methodology, including systematic investigation, systematic experimental design, systematic data storage (the base of systematic data management), and systematic data analysis and simulation.

## 3.2   A Multi-dimension Framework of Data-Brain

Based on systematic BI methodology, we design a multi-dimension framework of Data-Brain as shown in Fig. 1, including four dimensions and multiple conceptual views. We only give two conceptual views, the reasoning centric view and the computation centric view, as examples in Fig. 1 because of the limitation of space.

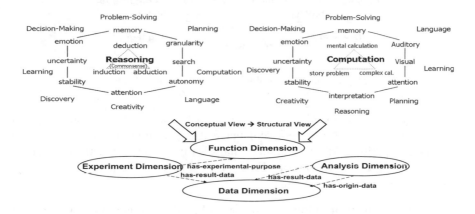

**Fig. 1.** A multi-dimension framework of Data-Brain

Firstly, the Data-Brain provides multiple conceptual views of systematic investigation of human thinking centric cognitive functions. The capabilities of human intelligence can be broadly divided into two main aspects: perception and thinking. Our BI study focuses on the "thinking centric" cognitive functions, which are complex and closely related to each other. Thus, we need to construct various conceptual views which illustrate systematic BI investigation and their inter-relationships from different viewpoints based on functional relationships among related human cognitive functions. These conceptual views can be regarded as cognitive/brain scientists' interfaces to facilitate their own research activities and cooperation with different focusing and research issues.

Figure 2 gives an abstract representation of the conceptual view, which illustrates reasoning centric, thinking oriented BI investigation and their inter-relationships based on functional relationships among related human cognitive functions. The core issue is to investigate human deduction, induction, and abduction related reasoning mechanisms, as well as including commonsense reasoning, as shown in the central of Fig. 2. Heuristic search, attention, emotion and memory are some component functions to implement human reasoning, as well as (information) granulation, autonomy, stability and uncertainty are some interesting characteristics, which need to be investigated with respect to human reasoning, as illustrated in the middle circle of this figure. Furthermore, decision-making, problem-solving, planning, computation, language, learning, discovery and creativity are the major human thinking related functions, which will be studied systematically, as illustrated outside the middle circle of this figure.

**Fig. 2.** The "Reasoning" centric conceptual view

Secondly, the conceptual view of the Data-Brain can be transformed into its own structural view with four dimensions, namely function dimension, data dimension, experiment dimension, and analysis dimension, respectively. Figure 1 illustrates such a transformation. Here we give more descriptions on the four dimensions as follows:

- The **function dimension** is a conceptual model of domain knowledge aiming at the systematic investigation in BI methodology. It describes the information processing courses of human thinking centric cognitive functions and functional relationships among them at the conceptual level. As stated above, the thinking centric cognitive functions are complex and closely related to each other, so Data-Brain needs to include a function dimension. The function dimension provides a holistic, conceptual functional model of human brain for systematic BI study. It also provides a machine-readable knowledge base for constructing various conceptual views.
- The **data dimension** is a conceptual model of domain knowledge aiming at systematic brain data storage. It describes multiple views, schemes, and organizations of human brain data with multiple data sources, multiple data forms, multiple levels of data granularity at the conceptual level. Conceptual modeling heterogeneous brain data using the data dimension is the key to realize systematic data storage which is the base of the systematic data management in BI methodology. By relations with the function dimension, the data dimension provides a conceptual data model which represents functional relationships among multiple human brain data sources obtained from various cognitive experiments with respect to all major aspects and capabilities of HIPS. This kind of data representation is with multi-level by modeling,

abstracting and transforming for multi-aspect analysis and simulation. The data dimension supports the implementation of a grid-based, analysis and simulation oriented, dynamic, spatial and multimedia database for storing and sharing the heterogeneous brain data efficiently and effectively.

- The **experiment dimension** is a conceptual model of domain knowledge aiming at the systematic experimental design in BI methodology. It describes characteristics of various experimentation plans, their classification and inter-relationships at the conceptual level. Systematic experimental design is an important issue of BI methodology. For uncovering the principles and mechanisms of HIPS, BI researchers need to design a series of cognitive experiments to obtain high quality of experimental data, which represent different aspects of various thinking centric cognitive functions, based on systematic methodology of cognitive experimental design. Thus, Data-Brain needs to include an experiment dimension. By relations with the function dimension and data dimension, the experiment dimension explicitly describes various relationships among various data sources. By the experiment dimension, cognitive experiments related information can be stamped on each data set for supporting systematic data analysis and simulation.

- The **analysis dimension** is a conceptual model of domain knowledge aiming at the systematic data analysis and simulation in BI methodology. It describes characteristics of various analysis and simulation methods, as well as their relationships with multiple human brain data for multi-aspect analysis and simulation. Agent-enriched data mining for multi-aspect data analysis is an important issue of BI methodology because the brain is too complex for a single data mining algorithm to analyze all the available data. Thus, Data-Brain needs to include an analysis dimension. Based on the analysis dimension corresponding to the data and experiment dimensions, various methods for data processing, mining, reasoning, and simulation can be deployed as agents on a multi-phase process for performing multi-aspect analysis as well as multi-level conceptual abstraction and learning, which aims at discovering useful knowledge to understand human intelligence in depth [12].

As shown in Fig. 1, these dimensions are not isolated from each other, but with various relations among them.

## 4   Data-Brain Modeling

### 4.1   Data-Brain Modeling Based on Brain Informatics Methodology

As a network/grid level of conceptual data model, Data-Brain modeling can adopt an ontological modeling approach. Comparing with other ontology engineering methodologies, the "Enterprise Methodology" [8] especially focuses on the construction of concept hierarchy and more fit to construct a Data-Brain which take concept hierarchies as skeletons of dimensions. Thus, we choose it to construct a Data-Brain. The course mainly includes the following four steps:

- defining the domain and scope of a Data-Brain
- identifying key concepts and properties
- defining the concept hierarchy through taxonomic relations
- constructing axioms

All of the above work, from the identification of key concepts to the classification of concepts, needs to obey some rules. During Data-Brain modeling, these rules come from systematic BI methodology, i.e., Data-Brain modeling is based on BI methodology.

According to dimension definitions, different dimensions of a Data-Brain focus on different aspects of data related domain knowledge. Thus, after defining BI as the domain of Data-Brain and the domain knowledge about BI data as the scope of Data-Brain, we can follow the remanent three steps of "Enterprise Methodology" to respectively construct four dimensions of Data-Brain based on the different issues of BI methodology, including:

- constructing the function dimension based on the systematic investigation
- constructing the data dimension based on the systematic data storage
- constructing the experiment dimension based on the systematic experimental design
- constructing the analysis dimension based on the systematic data analysis and simulation.

In the next section, we will describe how to construct the function dimension using OWL-DL language [19] and protege tool [20], which can be regarded as an example of the BI methodology based dimension construction.

## 4.2   Constructing the Function Dimension Based on Systematic Investigation

The function dimension constructed by ontology is an ontology of human cognitive functions. At present, most of human brain related ontologies are the brain structure related ontologies, such as the ontology of brain cortex anatomy [3]. The study about the ontology of cognitive functions has not gotten attention.

The systematic investigation is an important issue in BI methodology. Based on it, the course of function dimension construction can be described as follows:

- **Identifying key concepts and properties**: The systematic investigation in BI methodology is the investigation of human thinking centric mechanisms. Thus, key concepts of the function dimension are human cognitive function concepts, such as "Reasoning" and "Problem-solving", and their sub-function concepts, such as "Induction". These key concepts are described by properties, including data properties and object properties. The data properties are used to describe concepts themselves; the object properties are used to describe relationships among concepts. In this systematic investigation, each cognitive function concept represents a series of study activities which aim at the cognitive function. So it is worthless to describe cognitive function concepts themselves in the function dimension. For providing a holistic study view and functional model of human brain, the function dimension needs to focus on functional relationships among cognitive functions. Thus, there is no key data property in the function dimension. Only the key object property "has-functional-relationship-with", which describes functional relationships among cognitive functions, is included in function dimension. It includes various sub-properties, such as "includes-in-function" and "related-to-in-function", which are used to describe different functional relationships.

- **Defining the concept hierarchy**: At present, there is not standard taxonomy of human cognitive functions. Researchers often classify cognitive functions according to their own study viewpoints, such as LRMB model [10]. The systematic investigation in BI methodology is a thinking centric one, so we can define the concept hierarchy of function dimension as follows. Firstly, human cognitive function concepts are classified into two classes, "Perception-Centric-Cognitive-Function" and "Thinking-Centric-Cognitive-Function". The former includes perception oriented cognitive functions related concepts, such as "Vision" and "Hearing". The latter includes thinking centric cognitive functions related concepts on which BI focuses, such as "Reasoning". Secondly, all of cognitive functions are specialized into more characterized sub-classes. For example, the concept "Reasoning" can be specialized into multiple sub-concepts, such as "Induction" and "Deduction".

- **Constructing axioms**: Axioms are formal assertions that model sentences that are always true. They provide a way of representing more information about concepts, such as constraining on their own internal structure and mutual relationships. The primary axioms in Data-Brain are restriction axioms, including value constraints and cardinality constraints. Thus, constructing axioms in Data-Brain can be specialized as using data properties and object properties to describe concepts with the necessary constraints. Because of lacking data properties, constructing axioms in function dimension is just to use the key object property "has-functional-relationship-with" and its sub-properties to describe cognitive function concepts of function dimension with the necessary constraints. For example, we can use the object property "includes-in-function" to describe the concept "Induction" as follow:

$$Induction \subseteq Restriction(\exists includes-in-function\ Attention).$$

This means that *Induction* includes *Attention* as a sub-component, but it does not only include *Attention*.

## 4.3   Constructing Conceptual Views

A Data-Brain includes various conceptual views which are extracted from the function dimension of the Data-Brain. Because we use ontologies to model Data-Brain, its conceptual views are just the traversal views [7] of the function dimension and can be defined based on the definition of traversal view. Firstly, we give some related definitions:

**Definition 1.** *A* **View Core**, *denoted by Core, is a thinking centric cognitive function concept in the function dimension. As the core of a conceptual view, it is corresponding to a BI study issue and represents all the study activities of this issue.*

**Definition 2.** *A* **Traversal Directive** *for the source ontology O, denoted by TD, is a pair:*

$$< C_{st}, RT >,$$

*where $C_{st}$ is a concept in the source ontology O from which view is extracted, and represents the starter concept of the traversal; $RT =< R, n >$ is a relation directive, where R is a relation in O and n is a nonnegative integer or infinity which specifies the depth of the traversal along the relation R. If $n = \infty$, then the traversal includes a transitive closure for R starting with $C_{st}$.*

**Definition 3.** *A* **Traversal Directive Result** *(the result of applying directive TD to O), denoted by $TD(O)$, is a set of concepts from the source ontology O such that:*

1. *$TD =< C_{st}, RT >$;*
2. *$RT =< R, n >$, $n > 0$. if $C_{st}$ is a concept in the starts of the relation R, and a concept $C \in O$ is the corresponding end of the relation R, then C is in $TD(O)$;*
3. *$RT_{next} =< R, n-1 >$ is a relation directive. if $n = \infty$, then $n - 1 = \infty$. For each concept F that was added to $TD(O)$ in step 2, the traversal directive result $TD_F(O)$ for a traversal directive $TD_F =< F, RT_{next} >$ is in $TD(O)$.*

*No other concepts are in $TD(O)$.*

Based on the above definitions, the conceptual view of a Data-Brain can be defined as follows:

**Definition 4.** *A* **Conceptual View**, *denoted by CV, is a five-tuple:*

$$(Core, CF, CFIC, RF, R),$$

*where*

- *$CF =< Core, < parentClassOf, \infty >> (FD)$ is a specialization of $TD(O)$ and represents the concept set of core cognitive functions in a conceptual view, where "parentClassOf" is the inverse relation of the relation "subClassOf", and FD is the function dimension of a Data-Brain;*
- *$CFIC =< Core, < includes-in-function, \infty >> (FD)$ is a specialization of $TD(O)$ and represents the concept set of component cognitive functions and interesting characteristics in a conceptual view, where "includes-infunction" is a relation in the function dimension and used to describe the functional part-whole relationship among cognitive functions;*

---

**Algorithm 1.** Conceptual View Construction

---

**Input:** $Core$ and $FD$.
**Output:** $CV$.
1. Initialize empty concept sets $CV.CF$, $CV.CFIC$ and $CV.RF$;
2. Set $CV.Core = Core$;
3. Set $CV.R = \{$"$parentClassOf$", "$includes\text{-}in\text{-}function$", "$related\text{-}to\text{-}in\text{-}function$"$\}$;
4. $CV.CF = TDR(Core, "parentClassOf", \infty, FD)$;
5. $CV.CFIC = TDR(Core, "includes\text{-}in\text{-}function", \infty, FD)$;
6. $CV.RF = TDR(Core, "related\text{-}to\text{-}in\text{-}function", \infty, FD)$;
7. return $CV$

---

**Algorithm 2.** Getting Traversal Directive Result: TDR

---

**Input:** $C_{st}$, $R$, $n$, and $O$.
**Output:** $Concepts$.
1. Initialize an empty set of result concepts, $Concepts$;
2. Initialize the depth of the traversal, $depth = n$;
3. If ($depth == 0$) then
4.    return $Concepts$;
5. $depth_{next} = depth$;
6. If ($depth_{next} <> \infty$) then
7.    $depth_{next} = depth_{next} - 1$;
8. If ($R == "parentClassOf"$) then
9.    For each class $c_i$ in $O$
10.      If ($c_i$ $subClassOf$ $C_{st}$) then
11.        Add $c_i$ into $Concepts$;
12.        Add $TDR(c_i, R, depth_{next}, O)$ into $Concepts$;
13.      End If
14.    End For
15.Else
16.    For each $Restriction$ in $C_{st}$
17.      If ($Restriction$ is a value constraint $and$ its property name $== R$)
         then
18.        $c_i =$ Range of $Restriction$;
19.        Add $c_i$ into $Concepts$;
20.        Add $TDR(c_i, R, depth_{next}, O)$ into $Concepts$;
21.      End If
22.    End For
23.End If
24.return $Concepts$

---

- $RF =< Core, < related\text{-}to\text{-}in\text{-}function, \infty >> (FD)$ *is a specialization of* $TD(O)$ *and represents the concept set of related cognitive functions in a conceptual view, where "related-to-in-function" is a relation in the function dimension and used to describe the functional pertinence among cognitive functions*

- $R = \{parentClassOf, includes\text{-}in\text{-}function, related\text{-}to\text{-}in\text{-}function\}$ is a set of relations which are used to construct the conceptual view.

According to the above definitions, the algorithm for extracting a conceptual view from an OWL-DL based Data-Brain is shown in Algorithm 1. Its input parameters are the view core $Core$ and the function dimension $FD$. Its output is the $Core$ centric conceptual view $CV$. The function $TDR$ shown in Algorithm 2 is used to get the traversal directive result. Its input parameters are $C_{st}$, $R$, $n$, and $O$, which are corresponding to the starter concept, the name of relation, the depth of the traversal, and the source ontology of the definition of traversal directive result, respectively.

Using the above algorithms, we can choose different cognitive function concepts as view cores to construct various conceptual views based on various viewpoints of BI investigation.

## 5    Conclusions

The Data-Brain modeling is a core issue of BI study. Aiming at the requirements of systematic data management, this paper proposes a new conceptual data model of brain data, called Data-Brain. By the BI methodology based approach of Data-Brain modeling, a multi-dimension, formal Data-Brain can be constructed. It will be the core component of BI portal and play an important role in systematic BI study by providing the following functions:

- A formal conceptual model of human brain data, which explicitly describes the relationships among multiple human brain data, with respect to all major aspects and capabilities of HIPS, for supporting systematic human brain data storage, integration and sharing;
- A knowledge-base of systematic BI study, which stores brain data related multi-aspect domain knowledge to both support various knowledge driven data applications and provide valuable knowledge sources for solving special domain problems, such as the diagnosis and treatment for MCI (mild cognitive impairment) patients;
- A global view and knowledge framework for constructing a BI portal, on which various methods for data processing, mining, reasoning, and simulation are deployed as agents for implementing the Data-Brain driven multi-aspect data analysis.

## Acknowledgments

The work is partially supported by National Natural Science Foundation of China (No. 60673015), the grant-in-aid for scientific research (No. 18300053) from the Japanese Society for the Promotion of Science, and Support Center for Advanced Telecommunications Technology Research, Foundation.

# References

1. Chen, P.: The Entity-Relationship Model-towards a Unified View of Data. ACM Transactions on Database Systems 1(1), 9–36 (1976)
2. Chen, J.H., Zhong, N.: Data-Brain Modeling Based on Brain Informatics Methodology. In: 2008 IEEE/WIC/ACM International Conference on Web Intelligence (WI 2008), pp. 41–47. IEEE Computer Society Press, Los Alamitos (2008)
3. Dameron, O., Gibaud, B., et al.: Towards a Sharable Numeric and Symbolic Knowledge Base on Cerebral Cortex Anatomy: Lessons from a Prototype. In: AMIA Symposium (AMIA 2002), pp. 185–189 (2002)
4. Fonseca, F., Martin, J.: Learning the Differences between Ontologies and Conceptual Schemes through Ontology-Driven Information Systems. JAIS, Special Issue on Ontologies in the Context of IS 8(2), 129–142 (2007)
5. Jarrar, M., Demey, J., Meersman, R.: On Using Conceptual Data Modeling for Ontology Engineering. In: Aberer, K., March, S., Spaccapietra, S. (eds.) Journal of Data Semantics. LNCS, vol. 2800, pp. 185–207. Springer, Heidelberg (2003)
6. Jin, H., Sun, A., et al.: Ontology-based Semantic Integration Scheme for Medical Image Grid. International Journal of Grid and Utility Computing 1(2), 86–97 (2009)
7. Noy, N.F., Musen, M.A.: Specifying Ontology Views by Traversal. In: McIlraith, S.A., et al. (eds.) ISWC 2004. LNCS, vol. 3298, pp. 713–725. Springer, Heidelberg (2004)
8. Uschold, M., King, M.: Towards Methodology for Building Ontologies. In: Workshop on Basic Ontological Issues in Knowledge Sharing, held in conjunction with IJCAI 1995 (1995)
9. Uschold, M., Gruninger, M.: Ontologies Principles, Methods and Applications. Knowledge Engineering Review 11(2), 93–155 (1996)
10. Wang, Y.X., Wang, Y., et al.: A Layered Reference Model of the Brain (LRMB). IEEE Transactions on Systems, Man, and Cybernetics (C) 36, 124–133 (2006)
11. Zhong, N., Liu, J., Yao, Y.Y.: In Search of the Wisdom Web. IEEE Computer 35(11), 27–31 (2002)
12. Zhong, N.: Building a Brain-Informatics Portal on the Wisdom Web with a Multi-layer Grid: A New Challenge for Web Intelligence Research. In: Torra, V., Narukawa, Y., Miyamoto, S. (eds.) MDAI 2005. LNCS (LNAI), vol. 3558, pp. 24–35. Springer, Heidelberg (2005)
13. Zhong, N.: Impending Brain Informatics (BI) Research from Web Intelligence (WI) Perspective. International Journal of Information Technology and Decision Making 5(4), 713–727 (2006)
14. Zhong, N.: Actionable Knowledge Discovery: A Brain Informatics Perspective. Special Trends and Controversies department on Domain-Driven, Actionable Knowledge Discovery, IEEE Intelligent Systems, 85–86 (2007)
15. Australian EEG Database, http://eeg.newcastle.edu.au/inquiry/
16. Simulated Brain Database, http://www.bic.mni.mcgill.ca/brainweb/
17. Brain Bank, http://www.brainbank.cn/
18. The fMRI Data Center, http://www.fmridc.org/f/fmridc
19. http://www.w3.org/2004/OWL/
20. http://protege.stanford.edu/
21. Olfactory Receptor DataBase, http://senselab.med.yale.edu/ORDB/default.asp

# Combine the Objective Features with the Subjective Feelings in Personal Multi-alternative Decision Making Modeling

Jiyun Li[1] and Jerome R. Busemeyer[2]

[1] School of Computer Science and Technology, Donghua University,
No. 1882, West Yan'an Road, Shanghai, the People's Republic of China, 200051
[2] Psychology and Brain Science Department, Indiana University, Bloomington, IN 47405
jyli@dhu.edu.cn, jbusemey@indiana.edu

**Abstract.** In this paper we propose a computer modeling framework for personal multi-alternative decision making. It integrates both the objective feature space searching and evaluation and subjective feeling space deliberating and evaluation process. A case study on fashion decision making is given out as an example. It shows that the proposed model outperforms the currently widely studied ones in terms of prediction accuracy due to the consideration of the stochastic characteristics and the psychological effects that occur quite often in human multi-alternative decision making.

## 1 Introduction

The kernel of a computer recommendation system or computer aided decision making system is the model of decision making and preferences of the decision maker. Most of the decision making processes in them can be taken as a multi-alternative decision making process. Economists use preference in rational decision making to interpret decision making activities in terms of preference, looking at the effects or influences of preferences in making reasoning faster and introducing new decisions. While the focus of economists on "nice" preferences satisfying the properties [1] such as Coherent, Consistent, Complete, Comparatively constant, Concrete, Conveniently cleaved, completely comparative, etc., many studies in behavioral psychology have suggested many irrational effects exist in human judgment and decision making practice. These are important factors for personal decision making modeling. A good computer model for personal decision making should both help the decision maker make quick and reasonable decision, and in the mean time, make the decision according to his 'own' taste. In order to achieve this goal, we should have a better understanding of human decision making process, especially the personal factors' part.

Generally speaking, decision making process can be divided into three processes, namely, search, evaluation and deliberation. Both the search and evaluation strategy and deliberation and evaluation process have been carefully studied separately all these years. Many models and algorithms have thrived out. In the searching field,

N. Zhong et al. (Eds.): BI 2009, LNAI 5819, pp. 194–202, 2009.

genetic algorithm [2], ant colony algorithm [3], etc. are just a few. For the deliberation part, ever since Von Neumann and Morgenstern's [4] classic expected utility theory, much development has popped out from both rational theorist and behavioral scientists [5, 6, 7, 8]. They tackle the decision making problem from the representation of the preference of an ideal decision maker and identification of the behavioral principles that human preferences actually obey, respectively. Busemeyer et al. [9] study decision making from a dynamic cognitive perspective. They paid much more attention to the dynamic motivational and cognitive process during decision making, thus their models are good candidates for both modeling and predicating personalized preference in decision making process especially under risky or uncertainty. Besides, the computational attribute of their model makes it a good candidate to be incorporated into other computer systems than the descriptive ones.

From computer decision making system's perspective, scientists have realized almost all the searching algorithms, and have incorporated the rational expected utility theory into their implemented decision support systems. This makes it possible to provide the general decision support by taking advantage of the large capacity and huge memory of the computer system. But for personal decision making, in which the behavioral research plays an important role, computer systems till now have done quite less except some tracking and simple statistics of the user usage information.

In this paper, we will propose a personal decision making model which incorporate both the objective feature space searching and subjective feeling deliberating into the decision making process. Section 2 is an overview of the architecture of the model and details the different components of the architecture. Section 3 gives a case study of the proposed architecture in personal fashion decision making system. Section 4 includes some discussions and points out further research directions.

## 2  Proposed Architecture

The many factors that affect the process and result of a decision making can be roughly divided into two categories: objective factors and subjective factors. For the objective part, we refer to those factors that are definite and easy to be defined. Due to its objective nature, they are easy to be formalized and be accepted by computer systems. We have a relatively better understanding of this part. Models based on these factors are widely studied and implemented in different areas; but for the subjective parts, which are     relatively less studied, and due to the stochasticity and randomness exist in them, as far as our knowledge is concerned, less computer decision making systems have taken much consideration in this aspect. But we would like to argue that the subjective experiences that accompany our thinking process are informative in their own right in decision making processes. We therefore can't understand and model human judgment and decision making without taking the interplay of these subjective and experiential information into account. But there are many subjective factors like motivation, feeling, etc. which will be effective in decision making, it is not easy or almost impossible for us to define which one will play a major role in a specific decision making process, let alone consider all of them in one decision making model.

One of the possible ways to define then model the subjective factors is to confine the subjective factors that will be incorporated into the model only to those that have direct effects to decision results and have a semantic relationship with the objective features.

Based on the discussion above, we propose a decision making model whose architecture integrates both the objective searching and subjective deliberating—the evaluation process is incorporated in each separate process. The framework of the model is shown in Fig. 1. It takes into consideration both the objective and subjective factors during decision making process.

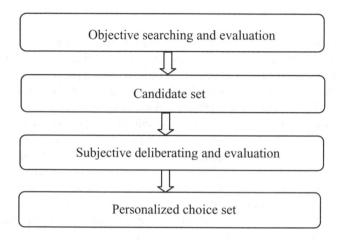

**Fig. 1.** Model architecture

We take the whole problem space $S_{whole}$ as two main spaces, one is the objective feature space $S_{objective}$, the other is the subjective feeling space $S_{subjective}$. The latter is intrigued from the former, thus can be inferred from the former by semantic mapping using semantic differential method proposed by Osgood [10].

$$S_{objective} \xrightarrow{\text{semantic mapping}} S_{subjective}$$

### 2.1 Objective Searching and Evaluation

At this stage, the model focuses on the objective feature factors of the problem space. These factors are easy to formalize and evaluate, and we can use genetic algorithm or other searching algorithms to accelerate the optimization process in decision candidate set finding.

### 2.2 Subjective Deliberating and Evaluation

During this stage, the main focus will be the subjective factors that have direct semantic relationship with the objective features in a decision making process. The goal is to approach the personal preference of the decision maker by modeling the deliberating process during decision making. It is mainly a trade-off between several conflict or incoherent subjective attributes in the subjective deliberating space. The model adopted here should be capable of accounting for the different 'irrational' phenomenon in human decision making behavior studies, so that the personalized choice set will be more human like. Here we choose Multi-alternative Decision Field Theory (MDFT) [11] that meets all these needs.

## 2.3  MDFT

MDFT is the expansion of DFT which was proposed by Jerome Busemeyer and Jim Townsend [9] in 1993. MDFT provides a mathematical foundation that leads to a dynamic, stochastic theory of decision behavior in an uncertain environment. According to MDFT, decisions are based on preference states, P(t), representing the momentary preference for choosing one alternative over the other in a binary choice or others in the ternary choices—for instance, A over B or A over B and A over C. That is, P(t) reflects the relative strength of preference of choosing A over B or A over B and C at time t. Preference evolves over time and is updated by an input valence, reflecting the momentary comparison of consequences produced by thinking of the different choices. The valence fluctuates because the decision maker's attention switches back and forth between anticipated consequences. The decision maker draws information about the alternatives from his or her memory. The possible consequences connected with either alternative are learned from experience and are remembered more or less well. A decision is made as soon as the preference process reaches a decision criterion or threshold for any one of the alternatives. The dynamics of the preference process is formally described according to the following stochastic process:

$$P(t + h) = SP(t) + V(t + h)  \tag{1}$$

Where,

P(t) is the preferences matrix of the alternative choices at time t;

S is the inhibition matrix. It is decided by the psychological distance of the two compared alternatives.

While

$$V(t) = C(t)M(t)W(t)  \tag{2}$$

C(t) is the contrast matrix, represents the process used to compare options;

For three alternatives,

$$C(t) = \begin{bmatrix} 1 & -0.5 & -0.5 \\ -0.5 & 1 & -0.5 \\ -0.5 & -0.5 & 1 \end{bmatrix}  \tag{3}$$

M(t) stands for the attributes' matrix in subjective deliberating space;

W(t) is the weight vector. It represents the amount of attention allocated to each attribute at each moment.

# 3   Case Study—Fashion Design Decision Making

## 3.1   Background

Intelligent fashion design system has been widely studied in the past decade. The main idea of the system is to take a piece of fashion or garment as a combination of several design items. So by searching from the database to find different items then combine, the system can generate many new style of garments. Thus it can help those who are not good at drawing to design their own fashion by making choices with aid by computer. The major problem of this kind of system is that fashion design is a creative and personal issue, there is not a general accepted rule set for the style. In order to tackle this problem, many researchers tend to adopt interactive genetic algorithm [12, 13, 14]. Many progresses have been made in capturing the user's personal style orientation. They can handle well the general rational fashion choice, but for the irrational phenomena that happen quite often in fashion choice or the stochastic characteristics of the user's choices, due to the limitation of GA, they failed.

Multi-alternative decision field theory models the dynamic, stochastic decision process in an uncertain environment. It can successfully accounts for many irrational behaviors in human multi-alternative decision making processes, such as similarity effect, attraction effect and comprise effect. But it suffers from that it usually allows less alternatives and attributes, and besides, the deliberation process is really time-consuming.

Thus it is reasonable to use the proposed framework, integrating GA searching algorithm with MDFT to have a better personal fashion decision making model.

## 3.2   Digitization

In order to ease the process of computer system, we digitize the original garment design picture into 10 design attributes, namely outline, waist line, length, collar, collar depth, segmentation, sleeve, sleeve length, ornaments, symmetry. They are used in GA searching. The 7 independent design items of them are used to find the semantic mapping relation of the objective feature space to the subjective feeling space. So a garment in the objective feature space will be expressed as a 7-tuple for semantic mapping:

$\text{Garment}_{oi}$ ( $\text{outline}_i$, $\text{waist}_i$, $\text{length}_i$, $\text{collar}_i$, $\text{sleeve}_i$, $\text{ornaments}_i$, $\text{symmetry}_i$ )

Where,

$\text{Outline}_i$, $\text{Waist}_i$ , $\text{Collar}_i$, $\text{Sleeve}_i$, $\text{Ornaments}_i$ each stands for the respective type of the $i_{th}$ garment;

$\text{Length}_i$ is the length of the garment;

$\text{Symmetry}_i$ stands for whether the garment is symmetry or not.

Fig. 2 is an example of the original garment design picture, table 1 gives out its digitization results. Where, A to G stand for the 7 design items, while the number follow them stands for the categories they fall into.

**Fig. 2.** Original bitmap

**Table 1.** Digitization and classification results of Fig. 2

| ID | Outline | Waist | length | collar | Sleeve | ornaments | symmetry |
|----|---------|-------|--------|--------|--------|-----------|----------|
| 10 | A1 | B2 | C2 | D1 | E2 | F2 | G1 |

### 3.3   The Objective Searching

We use genetic algorithm to search in the initial digitized 10 attributes objective feature space for garments with user designated style. For details of GA coding of the garment please refer to our former paper [15]. The fitness function of the genetic algorithm is approached by rules generated by data mining techniques from the judgments of fashion experts. The result we got will be the candidate set to be input into MDFT modeling.

### 3.4   Subjective Deliberating Modeling

In the subjective deliberating space, we are always facing with conflict feeling attributes. For example, we might want to be fashionable but in the mean time not too unique or we might want to be traditional but not too usual, etc. This is the main cause of the deliberating process in decision making. This kind of deliberation happens quite often when we are facing a relatively small set of candidate choices. We will deliberate and make some trade-off between some conflict attributes. MDFT can be used to model the trade-off deliberating process happened in decision making between two or more disaccord or even conflict attributes, but it is time consuming or almost impossible to deliberate among a large set of attributes, so we have to map the independent attributes in the high dimensional objective features space into the low dimensional negatively correlated subjective feeling space.

- Semantic mapping

Semantic mapping establishes relationship between customers' subjective feelings and the objective classification of design elements by mapping the perception of the clothing style to the semantic space that describes this perception.

Here we adopt the adjectives used to describe feelings for women's fashion collected by Ying Wang et al. [16], namely, simple/complicated, fashionable/traditional and unique/usual. By semantic mapping, a garment in subjective feeling space can be represented as a 3-tuple:

Garment$_{si}$ ( strucure$_i$, duration$_i$, acceptance$_i$ )

Where,

Structure$_i$ stands for the structure characteristics measurement of the $i_{th}$ garment ;

Duration$_i$ stands for the prevailing duration measurement of the $i_{th}$ garment;

And acceptance$_i$ stands for the acceptance degree measurement of the $i_{th}$ garment.

The relationship between the semantic space and the objective design space can be expressed as follows:

$$\text{Garment}_{si} = \text{Garment}_{oi} W \tag{4}$$

Where,

Garment$_{si}$ is a 1*3 vector. It stands for the 3 subjective measurements of $i_{th}$ garment in the subjective feeling space;

Garment$_{oi}$ is a 1*7 vector. It stands for the 7 objective features of $i_{th}$ garment in the objective feature space;

W is a 7*3 weight matrix, $w_{mn}$ stands for the $m_{th}$ objective feature's contribution to the $n_{th}$ subjective feeling evaluation. It was established by questionnaire survey and quantification analysis.

- Parameter fitting

By fitting the user's choice data into the MDFT model, we can get the individual weight vector for each of the factors in the subjective feeling space W$_{si}$.

The absolute value of each element in the inhibition S matrix is calculated by Euclidean distance between the two alternatives.

- Choice prediction

The stopping criterion for the MDFT is set as 500 time laps. After 500 times' loop, the one with the highest preference is chosen.

**Table 2.** Prediction accuracy contrast between GA+MDFT and IGA

| NO | GA+MDFT | IGA |
|----|---------|-----|
| 1 | 90% | 90% |
| 2 | 92% | 85% |
| 3 | 88% | 82% |
| 4 | 95% | 90% |
| 5 | 90% | 84% |
| 6 | 84% | 86% |

- Results

Right now, we have 6 subjects doing 100 trials. Each trial includes making ternary choice from 80 pieces of garments from the Database. Then we do the model fitting. Finally we use the parameters to predict the preference of the user for 80 garments generated by computer system. Experiments' results show that our model has a higher mean accurate prediction rate (90%) of the user's choice than the general IGA ones (86%).

## 4  Discussions and Conclusion

The model proposed in this paper incorporates both objective feature space searching and evaluation and subjective feeling space deliberating and evaluation. The case study on fashion decision making shows that it outperforms the generally used IGA. Through careful study of the decision process and the groups of garments for preference choice, we can find there are two reasons lead to the relatively poor performance of IGA.

- GA is based on a global search and global evaluation idea, while due to both the limitation of the human brain process and the computer screens, we only offer 3 each time, which means choice alternatives are mainly compared within 3 with some left memory of the earlier comparison result added to the final preference.
- The existence of those kinds of 'psychological effects' generally happened in multi-alternative human decision making, such as similarity effect, attraction effect and compromise effect, especially the similarity effect and compromise effect. Wherever these two kinds of effects exist in groups, the results will lead to the conflict with the IGA result, while MDFT is designed to explain and predict these effects.

To sum up, by the integration of genetic searching algorithm with MDFT, we can successfully account for both the decision time and accuracy of the computer based personal decision making process in the mean time.

Compared with the general decision making system, by incorporating the subjective feeling part using MDFT, we can:

- Clear identify the subjective factors—better understanding of the decision process, and will lead to a scalable and adjustable decision making model.
- Account for several classical irrational effects.-similarity effect, attraction effect and compromise effect, thus lead to a decision result that is more human like and can predict more accurately.
- Less interaction. –after model fitting, the proposed architecture needs less human interaction, than the IGA based intelligent fashion design system.

Compared with the mere cognitive dynamic decision making models for multi-alternative decision making, like MDFT, instead of using simple heuristics like eliminate by aspects, we introduced the optimized searching strategy to help quickly find the optimal solution in a larger objective feature space.

In the future study, we are intending to design other methodologies to find out or collect the subjective feeling factors, trying other dynamic cognitive models and also incorporate the user's feedback into the objective feature space evaluation and subjective feeling evaluation to have an efficient and more accurate personalized decision making model.

# References

1. Doyle, J.: Prospects for the Preferences. Computational Intelligence 20(2), 111–136 (2004)
2. Holand, J.: Adaptation in Natural and Artificial Systems. In: An Introductory Analysis with Applications to Biology,Control and Artificial Intelligence. University of Michigan Press, Ann Harbor (1975)
3. Dorigo, M.: Ant colony system-a cooperative learning approach to the traveling salesman problem. IEEE Transactions on Evolutionary Computation 1(1), 53–66 (1997)
4. Von Neumann, J.: Theory of games and economic behavior. Princeton University Press, Princeton (1947)
5. Janis, I.L., Mann, L.: Decision making: a psychological analysis of conflict, chlice and commitment. Free Press, New York (1977)
6. Kahneman, D., Tversky, A.: Prospect theory: an analysis of decision under risk. Econometrica 47, 263–291 (1979)
7. Machina, M.J.: Expected Utility analysis without the independence axiom. Econometrica, 277–323 (1982)
8. Wakker, P.P.: Additive representations of preferences. Kluwer Academic Publishers, Dordrecht (1989a)
9. Busemeyer, J.R., Townsend, J.T.: Decision Field Theory: A dynamic-cognitive aproach to decision making in an uncertain environment. Psychological Review 100(3), 432–459 (1993)
10. Simon, T.W., Schuttet, J.E., Axelsson, J.R.C., Mitsuo, N.: Concept, methods and tools in Sensory Engineering. Theoretical Issues in Ergonomics Science (5), 214–231 (2004)
11. Roe, R., Busemeyer, J.R., Townsend, J.T.: Multi-alternative decision filed theory:A dynamic connectionist model of decision making. Psychology Review 108, 370–392 (2001)
12. Nakanishi, Y.: Capturing preference into a function using interactions with a manual evolutionary design aid system. Genetic Programming, 133–138 (1996)
13. Lee, J.-Y., Cho, S.-B.: Interactive genetic algorithm for content-based image retrieval. In: Proceedings of Asia Fuzzy Systems Symposium, pp. 479–484 (1998)
14. Kim, H.-S., Cho, S.-B.: Application of interactive genetic algorithm to fashion design. Engineering Application of Artificial Intelligence 13, 635–644 (2000)
15. Li, J.Y.: A Web-based Intelligent Fashion Design System. Jounal of Donghua University 23(1), 36–41 (2006)
16. Wang, Y., Chen, Y., Chen, Z.-g.: The sensory research on the style of women's overcoats. International Journal of Clothing Science and Technology 20(3), 174–183 (2008)

# Automatic and Semi-automatic Approaches for Selecting Prominent Spatial Filters of CSP in BCI Applications

Nakarin Suppakun and Songrit Maneewongvatana

Department of Computer Engineering, Faculty of Engineering,
King Mongkut's University of Technology Thonburi,
Bangkok, 10140, Thailand
nakarin_sup@hotmail.com, songrit@cpe.kmutt.ac.th

**Abstract.** *Common Spatial Patterns (CSP)* has become a popular approach for feature extraction in *Brain Computer Interface (BCI)* research. This is because it can provide a good discrimination between 2 motor imaginary tasks. In theory, the first and the last spatial filters from CSP should exhibit the most discriminate between 2 classes. But in practice, this is not always true, especially if either of these 2 filters emphasizes on a channel that has high variability of variance of sample matrices among trials. Such spatial filter is unstable on a single class, and thus it is not appropriate to use for discrimination. Furthermore, one or both of these 2 spatial filters may not localize the brain area that relates to motor imagery. The desired spatial filters may be at the second or at an even greater order of sorted eigenvalue. In this work, we propose to find an appropriate set of spatial filters of CSP projection matrix, which may provide higher classification accuracy than using just 2 peak spatial filters. We present 2 selection approaches to select the set of prominent spatial filters: the first one is automatic approach; the second one is semi-automatic approach requiring manual analysis by human. We assessed both of our approaches on the data sets from BCI Competition III and IV. The results show that both selection approaches can find the appropriated prominent spatial filters.

## 1 Introduction

Brain computer interface (BCI) is a technology that finds new communicative ways of using only the brain to command machines. An electrode cap is placed on the head of the user for measuring electroencephalography (EEG) signals. To command the machine, the user imagines the specific task such as limb movement, composing words. These tasks affect the pattern of EEG signal. Computer would detect and classify these patterns into different tasks in order to control the computer application (such as cursor movement), or control the machine (e.g. wheelchair). In recent years, Common Spatial Patterns (CSP) has become a popular technique for feature extraction in BCI research. This is due to its good feature extraction performance that can be easily discriminated even by a simple classifier. Koles et al. introduced CSP to analyze EEG signals for discriminating between normal and abnormal EEGs [1].

N. Zhong et al. (Eds.): BI 2009, LNAI 5819, pp. 203–213, 2009.

Müller-Gerking et al. then applied CSP for feature extraction in BCI application [2]. Since some spatial filters from CSP projection matrix are useful than others, they selected a subset of spatial filters consisting of the first and the last $m$ spatial filters, $2m$ filters altogether. They claimed that the classification accuracy is insensitive to the choice of $m$. The optimal value of $m$ for their purpose was around 2 or 3. Much of the later work in BCI followed their guideline and usually set $m$ to be around 1-5. For example, $m$ was set to 3 in [3]; to 2 in [4], [5]; to 1 in [6]). In our previous work [7], we found that this claim was not always valid and higher accuracy can be achieved when setting the value of $m$ to be greater than 5. Therefore, in this paper, we explore the possibility to search for a reasonably good value of $m$ in each individual classification. We developed 2 selection approaches for finding this parameter: automatic and semi-automatic selection approaches. The main idea is to sort spatial filters based on their usefulness and identify the value of $m$ based on the drop of usefulness of spatial filters in this sorted order. Both selection approaches were assessed by using the dataset 1 from BCI Competition IV [8, 9] and data IVa from BCI Competition III [10]. The results show that our selection approaches usually provide a better value of $m$ and the values are sometimes optimal.

This paper is organized as follows. Section 2 presents theoretical background of CSP. Section 3 describes our selection approaches. Section 4 discusses the detail of datasets used in our experiments. Section 5 provides the experimental results. Discussion of results is presented in section 6 and the conclusion is given in section 7.

## 2   Theoretical Background on CSP and Analysis

In section 2.1, we provide an introduction to Common Spatial Patterns. We explain our view of CSP and the idea that convinced us to design an approach to select a subset of the spatial filters in section 2.2.

### 2.1   Common Spatial Patterns (CSP)

CSP can be used as a feature extraction tool in two-class models. The main idea of CSP is to find the projection matrix that maximizes variance for one class while minimizes variance for the other class. Once data set is projected into this space, variance can be used as a feature in classification process.

Let $X$ denote an $N \times T$ matrix of band passed and centered signal samples of a trial where $N$ is the number of channels and $T$ is the number of samples. Centering can be done for each channel individually by subtracting the mean amplitude of the particular channel from the signal amplitude for all samples in that channel. In order to use CSP, several trials of both classes are needed.

The normalized spatial covariance of a trial $X$ can be calculated as

$$C = \frac{XX^T}{trace(XX^T)} . \tag{1}$$

Let $C_{avg1}$ and $C_{avg2}$ denote the averages of covariance matrices of trials in classes 1 and 2 respectively, then

$$C_{sum} = C_{avg1} + C_{avg2}. \qquad (2)$$

Then $C_{sum}$ can be factored as $V_{sum}D_{sum}V_{sum}^T$ by eigenvalue decomposition, $V$ is the eigenvectors matrix and $D$ is the eigenvalues matrix.

Denote the whitening $P$ matrix as

$$P = (\sqrt{D_{sum}^{-1}}) \cdot V_{sum}^T. \qquad (3)$$

Applying $P$ to Eq. 2, in order to make the left side of the equation equal to identity matrix $I$

$$PC_{sum}P^T = PC_{avg_1}P^T + PC_{avg_2}P^T. \qquad (4)$$

$PC_{avg_1}P^T$ and $PC_{avg_2}P^T$ can be factored by eigenvalue decomposition as $QE_1Q^T$, and $QE_2Q^T$ respectively. We arrange these matrices so that the diagonal elements of $E_1$ and $E_2$ are sorted in descending and ascending order, respectively.

$$I = QE_1Q^T + QE_2Q^T. \qquad (5)$$

Finally we obtain $N \times N$ CSP projection matrix $W$ as

$$W = Q^T P. \qquad (6)$$

We called each row of $W$ a *spatial filter*, and each column of $W^1$ a *spatial pattern*. Typically, only some spatial filters are selected. For a certain value $m$, we reserved the first $m$ and the last $m$ rows of $W$ ($2m$ rows altogether) and removed the remaining middle $N - 2m$ rows. The remaining rows form $W_r$ matrix with dimension $2m \times N$. We can apply this matrix to any sample matrix $X_{raw}$ (similar to $X$ but without centering):

$$Z = W_r \cdot X_{raw}. \qquad (7)$$

A feature vector $f = [f_1, \ldots, f_{2m}]^T$ can be defined based on log of variance ratio:

$$f_i = \log\left(\frac{\text{var}(Z_i)}{\sum_{j=1}^{2m}\text{var}(Z_j)}\right), \qquad (8)$$

where $\text{var}(Z_j)$ is the variance of samples in row $j$. This feature vector is ready to serve as an input for any classifier. In this paper, we use simple Linear Discriminant Analysis (LDA) as the classifier in evaluation process. However, other classifiers can be used.

## 2.2 Our Analysis on CSP

Let us observe the derivation of CSP projection matrix. It is based on the average of covariance matrices of trials of the same class. If there is only one trial for each class, there is only one covariance matrix and the average covariance matrix is itself. However, in BCI, many trials of each class are required to properly train the CSP. Thus, it is possible that the values in a covariance matrix of a trial differ significantly from the values in the covariance matrix of another trial in the same class and therefore differ from the values in the averaged covariance matrix. In other words, the averaging process removes some information about stability of each spatial filter from CSP projection matrix computation.

Recall that the main idea of CSP is to find a projection that maximizes variance of one class and minimizes variance of the other class. In 2 dimension cases, the projection along the first spatial filter maximizes the variance of the first class and minimizes the variance of the second class while the projection along the second spatial filter maximizes the variance of the second class and minimizes the variance of the first class. In general, spatial filters in CSP projection matrix gradually shift the variances of projected data of two classes. Hence, the first and last spatial filters are the most useful when applied to averaged signals. However, this may not be true for in each individual trial since the signals can significantly differ from the average. This is why more spatial filters are needed. Furthermore, we found that for these 2 spatial filters, one or both of them may not localize the brain area around the related motor imaginary task, while the desired spatial filters are something else.

To address such problem, the coefficient of determination ($r^2$) was considered. It takes into consideration the stability of data between trials. We have used this parameter in both of our selection approaches to determine whether all training trials can be discriminated well when projected onto the prominent spatial filter dimensions.

## 3  Proposed Methods

We propose 2 selection approaches to select prominent spatial filter set: the first approach is automatic selection; the second approach is semi-automatic selection. The main idea is to extend the selection of spatial filters into ones with less significant eigenvalue but have high stability. In section 3.3, we provide a working example of selecting the value of $m$ by both approaches.

### 3.1  Automatic Selection Approach

The detailed steps can be expressed as follows:

1. Adjusts the sample length of each trial. We assume that all trials have the same number of samples. If this is not the case, we trim out some less-important periods such as the beginning of imagery in order to make their lengths equal. Divides the trials into training and testing sets.
2. Trains the CSP using training sets and obtains the full projection matrix $W$ (size $N \times N$). Projects all training data with this projection matrix. Each row of the projected data, called *component*, is the linear combination of channels.

3. Calculates the power spectrum for each component for all trials. The spectrum length of each components is $L_s$ where $L_s =$ is the lowest power of 2 that is greater than or equal to $L$. Let $S_i$ denote the $N \times L_s$ power spectra matrix for trial $i$.
4. Calculates $r^2$ coefficients (coefficient of determination). This can be done independently for each element of power spectra matrix. For a specific row and column in power spectra matrix, let $x_1$ and $x_2$ be a set of elements of matrices $S_i$ for all trials $i$ that belong to classes 1 and 2, respectively. Let $n_1$ and $n_2$ be the number of trials for classes 1 and 2, respectively. The values of $r^2$ can be computed by the following formula.

$$r^2 = \left( \frac{\sqrt{n_1 \cdot n_2}}{n_1 + n_2} \cdot \frac{mean(x_2) - mean(x_1)}{std(x_1 \cup x_2)} \right)^2 \tag{9}$$

where $std$ is the standard deviation. The resulting $N \times L_s$ coefficient of determination matrix is denoted as $R^{square}$. Notice that only spectra samples in band pass frequency range are calculated. For those whose frequency range is outside the band pass range will not be considered and are set to be 0.
5. Finds the value of $m$. We apply the maximum function to each row in $R^{square}$ and obtain vector $Rmax$. This vector is further divided into 2 halves: $Rmax_1$ and $Rmax_2$. Then, we try to locate the most stable spatial filter in each half. For the first half, we locate the element index of $Rmax_1$ that has the highest value of $r^2$. We also do the same thing for second half. Let $imax_1$, $imax_2$ denote these indices. Now we want to find the set of prominent spatial filters which consists of the first and last $m$ spatial filters of projection matrix $W$. The value of $m$ is the larger of $imax_1$ and ($N$-$imax_2$+1).

### 3.2 Semi-automatic Selection Approach

1. Follows the steps 1-4 of automatic selection approach described in section 3.1. Also, obtains the vector $Rmax_1$ and $Rmax_2$.
2. Plots the values in $Rmax_1$ and $Rmax_2$ in separate graph. Sorts the values in descending order. The graphs should be arranged so that y-axis is $r^2$ values and x-axis is the element index of the vector.
3. Manually locates the sharp drop. The last vector index before the drop is $idrop_1$ for the first graph and $idrop_2$ for the second graph.
4. Finds the maximum (minimum) index value from the portion of the sorted list between the first and $idrop_1$ ($idrop_2$). This value is $imax_1$ ($imax_2$) in the first (second) graph. The value of $m$ can be derived in the same way as in the first approach.

### 3.3 A Working Example of Selection the Value of $m$

For automatic selection method, $m$ is obtained by comparing $imax_1$ and $imax_2$ which are the element index of first bar of these 2 graphs. Thus $m = \max(1, 118 - 118 + 1) = 1$. For semi-automatic selection approach, sharp drops occur between the oblique shaded and solid bars in figure 1. The left graph (sorted values of $Rmax_1$) shows that $idrop_1$ is 2. In this case, $imax_1$ happens to be the same as $idrop_1$. For the right graph

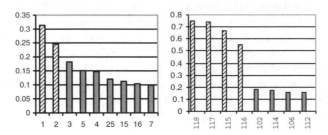

**Fig. 1.** Bar graphs show how semi-automatic selection works. $y$-axis represents $r^2$ value, and $x$-axis represents element index. (Number of channels, $N$, = 118).

($Rmax_2$), $idrop_2$ = 116 but the minimum element index is = 115 = $imax_2$. Thus $m$ = max(2, 118 – 115 + 1) = 4 as the result of semi-automatic selection approach.

## 4   Dataset Usage Detail

This section discusses on the usage detail for each dataset. We used 2 datasets: one from BCI Competition IV, dataset 1 [8, 9] and the other from BCI Competition III, dataset IVa [10]. Thus, we divide this section into 2 sub-sections corresponding to each dataset.

### 4.1   BCI Competition IV, Dataset 1

This dataset was provided by Blankertz, et al. [8], which recorded by [9]. However, this dataset consists of artificially generated EEG data, i.e. subjects $c$, $d$, and $e$. These subjects were excluded from our analysis. And the remaining subjects, i.e. $a$, $b$, $f$, and $g$ were used in our analysis. Subjects $a$ and $f$ imagined left hand vs. foot, subjects $b$ and $g$ imagined left hand and right hand. Details of experimental setup can be found on [8]. We used the downsampled version; 100 Hz, with 59 electrode channels.

We trained CSP and LDA on well-tuned window trials to continuously classify the continuous EEG data. Typically, signals are recorded between 0-4 seconds after the stimulus. But we adjusted the interval and starting time of each trial in order to achieve better performance. Relax trials that use to train a model are short idle periods between 2 consecutive motor imaginary periods. Starting relax trial was delayed after stop imagery for 1 second, to avoid responsive delay of subjects on stopping their imagination. Evaluation is based on classification result of each time sample.

### 4.2   BCI Competition III, Dataset IVa

This dataset was provided by Blankertz et al. [10]. Five healthy subjects, which consist of $aa$, $al$, $av$, $aw$, and $ay$. We have to discriminate between right hand and foot movement imageries. Details of experimental setup can be found on [10]. We used the downsampled version; 100 Hz, with 118 electrode channels.

For the given data, each subject has different number of training set and test set, we used first session as training set, and the remaining 3 sessions (thus, there are totally 4 sessions) were used as test set. This dataset was treated as single trial EEG for analysis. Trials were classified into motor imaginary of right hand vs. foot. Evaluation is based on classification result of each trial.

# 5  Experimental Results

This section showed results of our both selection approaches and the results of CSP using different values of $m$. The selection results of BCI Competition IV; dataset 1, can be seen on table 1, and BCI competition III; dataset IVa, can be seen on table 2.

**Table 1.** Selection results for value of $m$, testing on BCI Competition IV dataset1. There are 3 pairs of classes: task1, task2 and relax.

| Subject | Frequency and Time window | Between | Auto selection $m$ | Semi-auto selection $m$ |
|---|---|---|---|---|
| a | 7-28 Hz, 0.75-4.5 s | Task 1 vs. Task 2 | 2 | 2 |
|  |  | Task 1 vs. Relax | 3 | 3 |
|  |  | Task 2 vs. Relax | 24 | 24 |
| b | 11-14 Hz, 0.5-4.75 s | Task 1 vs. Task 2 | 1 | 1 |
|  |  | Task 1 vs. Relax | 3 | 3 |
|  |  | Task 2 vs. Relax | 1 | 1 |
| f | 7-26 Hz, 0.5-4.25 s | Task 1 vs. Task 2 | 2 | 3 |
|  |  | Task 1 vs. Relax | 3 | 4 |
|  |  | Task 2 vs. Relax | 3 | 4 |
| g | 7-26 Hz 0.5-4 s | Task 1 vs. Task 2 | 3 | 3 |
|  |  | Task 1 vs. Relax | 6 | 6 |
|  |  | Task 2 vs. Relax | 6 | 10 |

**Table 2.** Selection results for value of $m$, testing on BCI Competition III dataset IVa

| Subject | Frequency, Time window | Auto selection $m$ | Semi-auto selection $m$ |
|---|---|---|---|
| aa | 11-16 Hz, 0.5-3.5 s | 7 | 9 |
| al | 11-26 Hz, 0.75-4 s | 1 | 4 |
| av | 9-25 Hz, 0.75-3.5 s | 7 | 7 |
| aw | 10-24 Hz, 0.75-3.75 s | 1 | 3 |
| ay | 8-27 Hz, 0.25-3.5 s | 2 | 2 |

Figures 2 and 3 showed the plots of accuracy against the value of $m$ obtained from for the first and second datasets respectively.

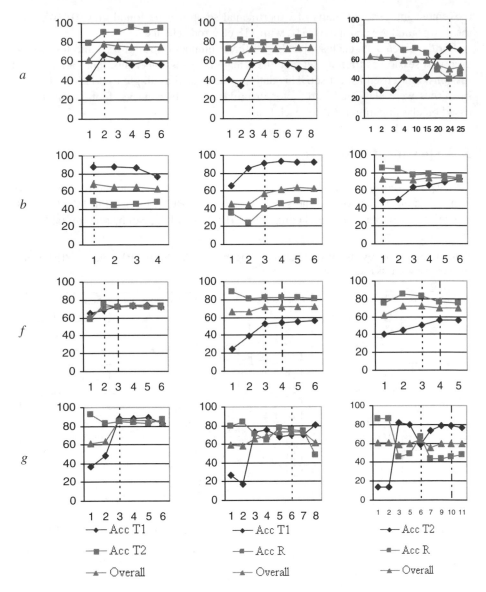

**Fig. 2.** Plotted results of the accuracy against *m* for each subject, testing on BCI competition IV; dataset 1. Here, y-axis is accuracy (%) and x-axis is *m*. Acc T1 is the percentage of correctly classifying Task 1. Acc T2 and Acc R are defined similarly. Dashed line represents the selected *m* using automatic approach, while dashed-dotted line represents the selected *m* using semi-automatic approach. If both approaches agrees on *m*, only one dashed line is shown. Each column shows results of the classification between a pair of tasks.

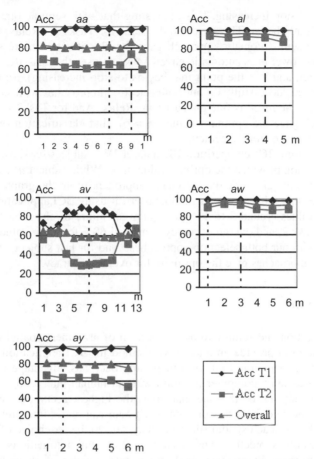

**Fig. 3.** Plotted results of accuracy against $m$ for each subject, testing on BCI competition III; dataset IVa. Acc T1 is the percentage of correctly classifying Task 1. Acc T2 is defined similarly. Dashed line represents the selected $m$ using automatic approach, while dashed-dotted line represents the selected $m$ using semi-automatic approach. If both approaches agree on $m$, only one dashed line is used for representation.

## 6  Discussion

For dataset from BCI competition IV; dataset 1, subject $a$ showed that both of our selection approaches gave the same values of $m$. Two out of three of binary classifications, Task1 vs. Task2 and Task1 vs. Relax, showed validity of our approaches. For the Task2 vs. Relax, CSP fails to provide suitable features. Regardless of the value of $m$, either one of two classes was classified with accuracy below a pure chance (50%). For subject $b$, our both selection approaches also gave the same values for all pairs, and the results are satisfied. The given parameters gave the overall accuracy on position that beginning to stop increasing when $m$ is increased. For subject $f$, on Task1 vs. Task2, the automatic approach output $m = 2$. This value yields a high overall accuracy which begins to stop increasing. It is interesting that with semi-automatic approach, accuracy is slightly better. This was also the same for Task1 vs. Relax. For Task2 vs. Relax, $m$ from semi-automatic approach gave the classification more stable in both classes than the lower $m$ as automatic gave (As we can see that classification of Task2 is slightly better, and it is the point that begins to stop increasing). Subject $g$, Task1 vs. Task2 showed successfully on our both selection approaches with same parameter given as $m = 3$. The result is fine on Task1 vs. Relax. And for Task2 vs. Relax, automatic approach gave the optimum value ($m = 6$), that classification of all aspect is stable, with the best overall accuracy.

For dataset from BCI competition III; dataset IVa, subject $aa$ showed that semi-automatic selection provided the optimal value of $m$. While subject $al$ was best at $m = 1$ which is the value given by the automatic approach. From the graph result of subject $av$, showed this subject is not a suitable user for using BCI application. As we can see from the values around 3 to 11, those showed failure of classifier that failed for discriminate Task 2, and overall accuracy is around 60% for all $m$ values. Subject $aw$ worked well with our both selection approaches. And for last subject, subject $ay$, our both selection approaches give the optimum for overall accuracy.

## 7 Conclusion

Automatic selection and semi-automatic selection of prominent spatial filters of CSP based on value of $m$ provide an alternative method for BCI application when CSP is used as a feature extraction tool. Automatic approach has advantage that the whole process can be done by computer program without human intervention. However, in some cases, it may not be the one that gives the highest accuracy. Semi-automatic approach can alleviate from this drawback. Human can better interpret the relativity between $r^2$ values of each spatial filter and may select a better choice of $m$. However, the selection can be subjective. The disadvantage of semi-automatic is of courses the needs human to manually analyze the dropping rate, which may require some experience person to analyze them effectively. Analyzing of dropping rate may be hard to determine by computer due to unpredictability on speed of dropping values for each subject, and each pair of 2 classes. In our opinion, semi-automatic can be converted to full automatic, if an efficient algorithm on analyzing of dropping rate was designed.

**Acknowledgments.** We would like to thank Benjamin Blankertz, Guido Dornhege, Matthias Krauledat, Klaus-Robert Müller, and Gabriel Curio for providing their datasets in our analysis.

## References

1. Koles, Z.J., Lazar, M.S., Zhou, S.Z.: Spatial Patterns Underlying Population Differences in the Background EEG. Brain Topography 2(4), 275–284 (1990)
2. Müller-Gerking, J., Pfurtscheller, G., Flyvbjerg, H.: Designing Optimal Spatial Filters for Single-Trial EEG Classification in a Movement Task. Clinical Neurophysiology 110, 787–798 (1999)

3. Blankertz, B., Tomioka, R., Lemm, S., Kawanabe, M., Müller, K.-R.: Optimizing Spatial Filters for Robust EEG Single-Trial Analysis. IEEE Signal Proc. Magazine 25(1), 41–56 (2008)
4. Ramoser, H., Müller-Gerking, J., Pfurtscheller, G.: Optimal Spatial Filtering of Single Trial EEG During Imagined Hand Movement. IEEE Trans. Rehab. Eng. 8(4), 441–446 (2000)
5. DaSalla, C.S., Kambara, H., Sato, M.: Spatial Filtering and Single-Trial Classification of EEG during Vowel Speech Imagery. In: 3$^{rd}$ International Convention on Rehabilitation Engineering & Assistive Technology. ACM Digital Library, New York (2009)
6. Wang, Y., Gao, S., Gao, X.: Common Spatial Pattern Method for Channel Selection in Motor Imagery Based Brain-computer Interface. In: IEEE Engineering in Medicine and Biology Society 27th Annual Conference, pp. 5392–5395 (2005)
7. Suppakun, N., Maneewongvatana, S.: Improving Performance of Asynchronous BCI by Using a Collection of Overlapping Sub Window Models. In: 3$^{rd}$ International Convention on Rehabilitation Engineering & Assistive Technology. ACM Digital Library, New York (2009)
8. Blankertz, B.: BCI Competition IV,
   http://ida.first.fraunhofer.de/projects/bci/competition_iv
9. Blankertz, B., Dornhege, G., Krauledat, M., Müller, K.-R., Curio, G.: The Non-invasive Berlin Brain-Computer Interface: Fast Acquisition of Effective Performance in Untrained Subjects. NeuroImage 37(2), 539–550 (2007)
10. Blankertz, B.: BCI Competition III,
    http://ida.first.fraunhofer.de/projects/bci/competition_iii

# Boosting Concept Discovery in Collective Intelligences

Francesca Arcelli Fontana[1], Ferrante Raffaele Formato[2], and Remo Pareschi[3]

[1] Dipartimento di Informatica, Sistemistica e Comunicazione.
Università di Milano " Bicocca"
[2] Dipartimento di Ingegneria Università del Sannio
[3] Dipartimento di Scienze e Tecnologie per l'Ambiente e il Territorio
Università del Molise

**Abstract.** Collective intelligence derives from the connection and the interaction of multiple, distributed, independent intelligent units via a network, such as, typically, a digital data network. As collective intelligences are effectively making their way into reality in consequence of ubiquitous digital communication, the opportunity and the challenge arise of extending their basic architecture in order to support higher thought-processes analogous to those characterizing human intelligences. We address here specifically the process of conceptual abstraction, namely the discovery of new concepts and ideas, and, to this purpose, we introduce the general functional notion of cognitive prosthesis supporting the implementation of a given thought-process in a collective intelligence. Since there exists a direct relationship between concept discovery and innovation in human intelligences, we point out how analogous innovation capabilities can now be supported for collective intelligences, with direct applications to Web-based innovation of products and services.

## 1 Introduction

The age of digital networking has turned into a real option the possibility of a collective intelligence [12], namely the coordinated activities of independent intelligent units into the activity of a larger intelligence, which finds its antecedents in a variety of contributions from visionary thinkers across human history, starting from Plato's panpsichic awareness to Bergson's and Deleuze's multiplicities and Verdnasky and Teilhard de Chardin's noosphere. Indeed, why then are computers and computer networks, such as Internet *in primis*, so crucial in order to give reality to an idea that, in various formats, has always been around? Precisely because they make it real by embodying it into an extremely efficient bio-digital machine, capable of connecting together multiple sources of intelligence through real-time interactions and of storing the results of such interactions in a long term memory. As long as this machinery was unavailable collective intelligence could not emerge as a new form of life.

Thus, the coming of age of collective intelligences can be viewed as an evolutionary milestone comparable, and to some extent analogous, to the appearance, roughly 2.5 billions of years ago, of modern eukaryotic cells, as found in most of the living

N. Zhong et al. (Eds.): BI 2009, LNAI 5819, pp. 214–224, 2009.

forms that are today inhabiting the earth, and as derived from ancient bacteria being engulfed by pre-eukaryotes through a form of endosymbiotic cooperation. Just as what is happening today, the availability of the appropriate biological infrastructure made possible, to our bacterial ancestors, a fundamental evolutionary leap, as it had not happened since the inception of life on Earth another billion years before and as may happen again today.

Our general approach is the implementation of specialized software layers   as kind of cognitive prostheses for the strong integration of distributed intelligence units in the performance of specific cognitive functions.   In the millions years of  history of the human brain, the same role is played by specialized regions supervising higher thought-processes. In that context, cognitive prostheses have been recently proposed and engineered by bio-medical researchers (see for instance [5]) as a way of restoring cognitive functions whose supervising regions have been damaged (as through the effect of anoxia ) or have decayed (as in the case of Alzheimer disease).  In the case of collective intelligence, our aim is to use them to implement higher thought-functions directly, thus supplementing  through technology what nature has accomplished in the case of human intelligence.

Our specific target  here pertains the activity of conceptual abstraction. Said in a more direct way, we want to support the processes of thought that get done whenever we utter, verbally or mentally, "I've got an idea!" or, to borrow the words of an ancient and most illustrious thinker and innovator, "Eureka!". Abstraction can then trigger innovation. Whenever this happens, something really new takes place, by combining the new idea with existing knowledge and skills, and thus bringing it back from the realm of abstraction into the real world where new things are created  that correspond to instances of the ideas. These can range, depending on the extent, the nature and the quality of the novelty, from new tricks to make ends meet and new ways to become wealthy by, say, manufacturing a type of ice-cream appealing to the population living north of the artic circle, to revolutionary outbreaks and timeless works of art. Of course, for this to happen, we  do not need just new ideas but also *good* new ideas, but the point here is that ideas, once generated, immediately undergo the process of natural selection and only the good ones make it back to the world by being instantiated into something that did not exist before. Now, it is a fact that ideas are generated not just by individuals but also by communities, which are themselves a fundamental ingredient of collective intelligences. The ideation capabilities of communities within corporate organizations have been widely studied, among others, by Nonaka and Takeuchi  in their book The Knowledge Creating Company [14] where they introduce an interesting construction called the "Knowledge Spiral" that relates the emergence of new ideas  ("Knowledge Externalization" in their terminology) with innovation. The Knowledge Spiral, illustrated in Figure 1, is composed of the following four phases linked together in a continuous cycle:

--Internalization, during which ideas that have been accepted as innovative get learned by people in the community that will use them to produce new artifacts, provide new services etc.;
--Socialization, during which the community develops a common understanding of the ideas thus internalized;

--Externalization, the crucial phase during which altogether new ideas emerge from the community;

--Combination, when new ideas are combined with older and established ones for innovation purposes.

On the other hand, while it is easy to identify a set of well-established technologies to support Internalization, Socialization and Combination of knowledge, including e-mail, corporate portals and groupware, so far nothing has been identified for Externalization. Generally speaking the attitude has been that we may hope that it will happen, and that the right kind of corporate culture and innovation will be responsible for creating favorable conditions for it. We want to show here that this does not have to be the case. In fact, without performing any kind of magic, our cognitive prosthesis will recognize ideas once they are likely to have been already produced in different parts of the community that underlies the collective intelligence itself. Indeed, in a collective intelligence new ideas are produced all the time, and they are immediately internalized by being freely exchanged. Once we identify strong internalization and socialization points that do not correspond to any so far known idea, then we have found new ideas that need to be accounted for. Before showing how this can be done we need, however, to extend the definition of community from its commonly understood meaning, and relate it directly to the collective intelligence *par excellence*, namely the Web. Furthermore, we need to define a minimal architecture for collective intelligences, where abstraction capabilities for representing ideas are made available.

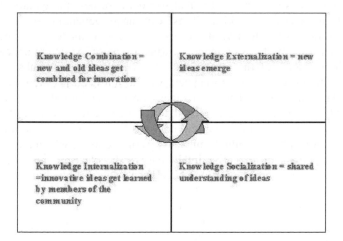

**Fig. 1.** The knowledge spiral

## 2  A Minimal Architecture for Collective Intelligences

At the basis of our minimal architecture there are ontologies and communities. Ontologies have since a long time provided a powerful tool to organize knowledge. At a philosophical level, ontology is the most fundamental branch of metaphisics.

It studies being or existence and its basic categories and relationships, to determine what entities and what type of entities exist. At the more specific level of knowledge representation and knowledge management, ontologies identify ideas as concepts applied to specific domains and organized as graphs via relationship links. A typical example of an ontology as shown in Figure 2 is given by an automotive ontology, organizing concepts used by enterprises operating in the automotive industry.

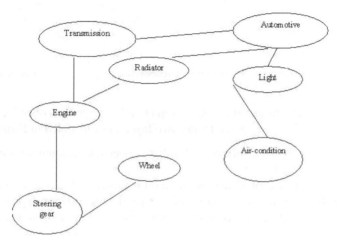

**Fig. 2.** An automotive ontology

Our view both of ontologies and of communities is information-driven: they are identified with the information they contain, either because they produce it (in the case of communities) or because they categorize it (in the case of ontologies). Furthermore, both communities and ontologies can be represented as networks (directed graphs). This common formal representation makes it easy to model the interaction between the two levels. Indeed, as a very important *caveat*, it should be made clear that the notion of community that we adopt here not only assumes networks as a form of representation, but is itself a specialization of the notion of network. In fact, we adhere to the view, coming from the tradition of network theory, that a community can be defined in topological terms as a region of a dynamic network where links are denser than in the surrounding regions. In other words, communities are directly identified with highly interconnected regions of dynamic networks, as shown in Figure 3. This allows us to model as communities social networks whose nodes map directly into human individuals, such as family clans, but also digital communities where the role of humans is crucial but indirect, in that the primary community members are Web sites pointing one to the other. As we shall discuss later on, the most immediate applications for our cognitive prosthesis supporting abstraction capabilities in collective intelligences are indeed in the domain of this kind of Web communities.

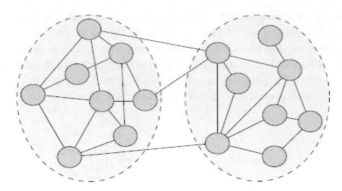

**Fig. 3.** Communities (Highly interconnected regions) within a network

Thus, we are ready to introduce our minimal architecture for collective intelligences, which defines structures given by just two layers, as illustrated in Figure 4:

- an "object" level with network of nodes that cluster into communities content and information;
- a "meta-level" where content and information is abstracted into ontologies and abstraction can be viewed also in terms of the symmetric relationship of classification/categorization, with concepts classifying the content they are abstractions of.

## 3   A Cognitive Prosthesis for Concept Discovery

The fundamental function that our cognitive prosthesis must supply is supporting abstraction into ideas. In other words, it must support concept discovery. Thus, our cognitive prosthesis will act as a  hardwired "knowledge spiral" that supervises the interaction between the two layers in the minimal architecture of collective intelligences and makes possible to up-raise the process of information creation at the object level into a process of knowledge creation at the meta-level through the introduction of emergent concepts into existing ontologies. Intuitively, the process that we apply is an interleaving of content-based classification together with graph-based detection of communities, where we assume that the existing nodes in the ontology classify corresponding community regions

- if the graph covered by the classification is part of still un-covered adjacent community graphs then we add corresponding nodes to the ontology, thus putting in action the learning capability to abstract from the new content which has just been acquired;
- we then proceed to classify again on the basis of the extended ontology.

Obviously, we also assume that the topology of the community structures is dynamic, and thus subject to continuous reconfiguration, since the participants in the collective

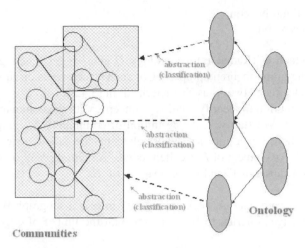

**Fig. 4.** A collective intelligence built up from a community layer

intelligence are constantly creating new information, and are connecting up with new nodes thus creating new communities.

In our model, we call *domain ontology* a directed graph $G = (C, R)$ where nodes are concepts and arches are binary relations. A *classifier* is a tuple $(TS, L, A, \varphi)$ such that:

- $TS$ is a training set of past cases
- $L$ is a set of labels
- $A$ is a set of actual cases
- $\varphi:A{\rightarrow}L$ is a *classification function.*

Let $l$ be a label in $L$. We call *seed relative to c* the set $\varphi^{-1}(c)$. Intuitively, in a Web context the seed relative to the concept c is the set of URL's that are definitely about $c$.

Given a domain ontology $O = (C, R)$, we perform *community trawling* upon $\varphi^{-1}(c)$ for each $c$ in $C$. Community trawling is a process that, starting from a graph G, finds all communities imbedded in the vicinity of G. See for example [11]. By doing so, for each concept $c \in C$, we obtain a set of graphs $C(c)$ that represents a bundle of Web communities pertaining to $c$.

**Example**

Let {www.autopartsfair.com,www.griffinrad.com,www.radiator.com,www.carpartswholesale.com} be the set of URLs of the example in Figure 2. We now apply the algorithm described in [11] to detect the Web communities subsumed by this set. The Web community associated with the concept "Radiator" is individuated by the following sites:

www.autoguide.net
www.autopartsfair.com
www.carpartswholesale.com

www.allworldautomotive.com
www.btztradeshows.com

By doing so, we attach $\varphi^{-1}$(Radiators) to ontology $O$ and we get a new graph as in Figure 6. In [1,2] the graph-theoretic properties of such structures is discussed, showing that they are scale-free networks as characterized in [3,4,5].

We are now in condition to identify ontology concepts with the communities originated by classification. We start with the definition of community . A *graph* G is an ordered pair $(V,E)$ where $V$ is a finite set whose elements are called nodes and $E$ is a binary relation on $V$. An element of $E$ is called edge. A *subgraph* W of G is a pair $(V',E')$ where V' is a subset of $V$ and $E'$ is a subset of $E$.

**Definition 1.** Let $G$ be a graph. A weak community in $G$ is a subgraph $W$ of $G$ such that the number of edges internal to $W$ is greater than the umber of edges external to $W$.

**Definition 2.** Let $G$ be a graph. A *strong community* in $G$ is a subgraph $W$ of $G$ such that, for any node in $W$, the number of edges internal to $W$ is greater than the number of edges external to $W$.

Of course, strong communities are also weak communities. Now we define the concept of smallest community generated by a graph.

**Definition 3.** Let $W$ be a subgraph of $G$. The smallest strong (weak) community generated by $G$ is a subgraph $C(W)$ of G such that
    i) $C(W)$ is a strong (weak) community
    ii) $C(W) \supseteq W$
    iii) If $C'$ is a community and $C' \supseteq W$ then $C' \supseteq C(W)$

The construction of $C(W)$ is straightforward and is given by the following algorithm

    1.   Input a graph $W=(V',E')$ of the graph $G=(V,E)$
    2.   for any node $v \in V'$
    3.   **if** the number of internal arches of n is greater than the number of its external arches
        begin

$$E_v = \{(v,a) \mid a \in V\}$$
$$V_v = \{a \in V \mid (v,a) \text{ is external to } W \}$$
$$V = V \cup V_v \quad R = R \cup R_v$$

       Go to 2
       end if

When the algorithm stops, the graph W will become $C(W)$. $C(W)$ is also called the *community generated by W*.

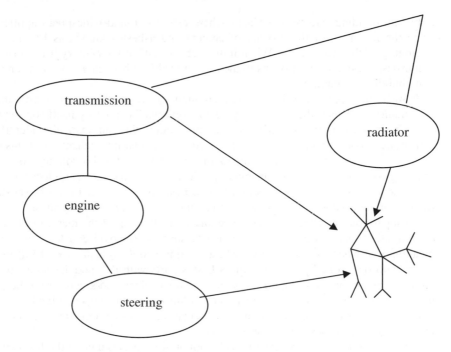

**Fig. 5.** Merging ontologies with communities

**Definition 4.** Let *S* be a seed and *IK(S)* be a community generated by *S*. We call *dynamic seed generated by S* the set of positive answers given by a classifier trained with S. We indicate this set with *Cl(IK(S)))*.

**Definition 5.** Let *S* be seed and let *Cl(IK(S)))* be the dynamic seed generated by S. We call dynamic concept generated by S, and we indicate it with Dyn(S), the web community dynamically generated by *Cl(IK(S)))*. Formally:

$$Dyn(S) = Kl(Cl(IK(S)))). \tag{1}$$

Whenever there is an evolution in the connectivity of *IK(S)*, two new seeds *S'* and *S''* are generated by *S*. By classifying *IK(S')* and *IK(S'')* with *S'* and *S''* as training sets, we get two new concepts c' and c'', respectively.

**Example 2 : Asbestos Removal meets Photovoltaic Roofs**
We simply illustrate the procedure above by applying it to a specific case related to a collective intelligence rooted in social networks focused on environmental issues. The process is graphically described in Figure 6.

Let us assume that our collective intelligence contains a concept corresponding to "Asbestos Removal" pointing to a specific community of Web sites. As is well-known, many buildings contain asbestos, which was used in spray-applied flame-retardant, thermal system insulation, and in a variety of other materials. Roofs of industrial buildings are typical places where asbestos found widespread application, since the invention of the mixture of cement and asbestos known as Eternit at the beginning of the last century. When in the second half of the century it became clear that asbestos can be very noxious to the human health, it became relevant to remove it from buildings containing it.

Now let us suppose that later on this community evolves by merging with another community for which the concept "Asbestos Removal" cannot by itself (on the basis of its training) classify properly. By our procedure, this automatically identifies a self-trained concept which we baptize (based on our own internalization of this community) "PV Integration", referring to the integration of photovoltaic modules in buildings so as to make them capable to produce, from the solar light, electricity to be consumed by the residential units in the buildings. For obvious reasons of exposure to light, roofs are indeed the most natural candidates for such an integration. We then add a super-concept which can be viewed as an abstraction of the merging of the two communities, and can be properly named "Asbestos Removal and PV-Integration", meaning that whenever we remove asbestos from a roof, we can as well kill two pigeons with one stone by replacing the asbestos parts with PV modules so as to make the building not only environmentally safe but also "clean and green" from the standpoint of energy production. The emergence of this concept is indeed at the basis of the integration, as is currently taking place, between the asbestos removal industry an the engineering of roof-based PV power plants.

The method for the discovery of Web communities is based upon the following result due to Ford and Fulkerson ([8]). We call s-t graph a structure of the form $(V,E,s,t)$.

**Theorem 1.** *Let $G = (V,E,s,t)$ be an s-t graph. Then the maximum flow is equal to the minimum s-t cut.*

A *minimum cut* is a cut which minimizes the total capacities of the cut edges and the following theorem, due to Ford and Fulkerson, states that the maximum graph flow equals to the value of a minimum cut.

**Theorem 2.** *Let $G=(V,E)$ be a graph. Then a community C can be identified by calculating the s-t minimum cut with s and t being used as the source and sink respectively.*

We then have the following algorithm of community extraction found by Flakes, Lawrence and Gilles.([7]). Suppose $T = V\backslash S$ for a subset $S \subseteq V$. If $s \in S$, $t \in T$ then the edge set $\{(u,v) \in E \mid u \in S, v \in T \}$ is called a *s-t cut* (or cut) on a graph $G$, and an edge included in the cut set is called a *cut edge*.

The procedure is as follows:

1. $S$ is a set of seed nodes and $G = (V,E)$ is a subgraph of the Web graph crawled within a certain depth from the nodes in $S$, i.e. certain in/out links away from the nodes in $S$. The subgraph $G$ is called *vicinity graph*.

2. Suppose that any edge $e \in E$ is undirected with edge capacity $c(e) = |S|$.

3. Add a virtual source node $s$ to $V$ with the edge connecting to all nodes in $S$ with edge capacity $= \infty$ and finally add a virtual sink node $t$ to $V$ with the edges connected from all the nodes in $V-\{S \cup \{s\} \cup \{t\}\}$ with the edge capacity $= 1$.

4.    Then perform the *s-t* maximum flow algorithm for $G$.

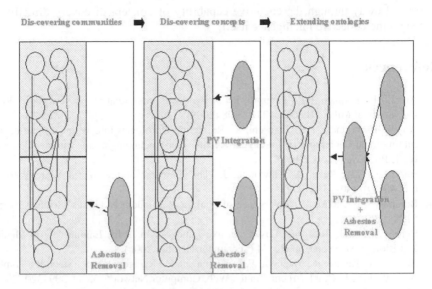

**Fig. 6.** Knowledge creation at work

## 4   Conclusions

We have formally defined the functioning of a cognitive prosthesis to boost concept discovery in collective intelligences, with the objective to transfer at the collective level the human capability to innovate. We are currently experimenting our approach in the perspective of meeting the marketing needs of corporations in the age of Web 2.0 and of empowered customer and user groups, as described in [13], and of supporting user-driven product innovation [9,10].

We conclude by replying to an obvious question: how does this fit with the Semantic Web project http://www.w3.org/2001/sw/, that, since its inception at the very beginning of the new millenium, aims at making the Web fully "understandable" by software agents through the annotation of Web content with ontologies?

By and large we view the Semantic Web as fitting with our vision and objectives, and we believe that its full realization will lead to the creation of a federation of advanced collective intelligences effectively marking the threshold for the Collective Intelligence era. We are proposing, however, an alternative to its current methods, which amount essentially to involving communities in the manual creation of meta-content in the form of ontologies and concept taxonomies. By contrast, we view communities as much better fitted at producing content rather than meta-content: this is the primeval impulse of human creativity that gets naturally boosted and multiplied through digital connectivity, as demonstrated by the explosive growth of Web 2.0 — namely the Web of blogs, social networks and personal spaces.

By providing a way to automatically co-evolve communities and ontologies, our framework can be exploited to automate the creation of a "meta-web" where the burden of semantic annotations is taken away from the users communities, and newly inserted concepts effectively correspond to genuine innovation, as captured and abstracted away through the cognitive capability of concept discovery directly supported by the collective intelligence itself.

# References

1. Arcelli, F., Formato, F.R., Pareschi, R.: Networks as interpretation of Network ch.Report University of Milano Bicocca –TD 23/08 (2008)
2. Arcelli, F., Formato, F.R., Pareschi, R.: Reflecting Ontologies into Web Communities. In: International Conference on Intelligent Agents. Web Technologies and Internet Commerce IEEE Press New York (2008)
3. Barabasi, A.L., Réka, A., Hawoong, J.: The diameter of the World Wide Web. Nature 401(9), 130–131 (1999)
4. Barabasi, A.L., Réka, A.: Emergence of Scaling in Random Networks. Science 10, 509–512 (1986)
5. Berger, T., et al.: Restoring Lost Cognitive Function. In: IEEE Engineering in Medicine and Biology Magazine, pp. 30–44 (September/October 2006)
6. Broder, A., Kumar, R., Maghoul, F.R., Raghavan, P., Rajagopalan, S., Stata, R., Tomkins, A.S., Wiener, J.: Graph structure in the Web. Computer Networks 33(1), 309–320 (2000)
7. Flake, G.W., Lawrence, S., Giles, C.L.: Efficient Identification of Web Communitie. In: Sixth ACM SIGKDD International conference on knowledge discovery and data mining, pp. 150–160 (2000)
8. Ford, L.R., Fulkerson, D.R.: Maximal flow through a Network. Canadian Journal of Mathematics 8, 399–404 (1956)
9. von Hippel, E.: The Sources of Innovation. Oxford University Press, Oxford (1988)
10. von Hippel, E.: Democratizing Innovation. MIT Press, Cambridge (2005)
11. Imafuji, N., Kitsuregawa, M.: Finding a web community by maximum flow algorithm with HITSscore based capacity. In: 8$^{th}$ Eighth International Conference on Database Systems for Advanced Applications, pp. 101–106 (2003)
12. Levy, J.P.: Collective Intelligence: Mankind's Emerging World in Cyberspace Helix Books (1998)
13. Li, C., Bernoff, J.: Groundswell: Winning in a World Transformed by Social Technologies Harvard Business Press (2008)
14. Nonaka, I., Takeuchi, F.: The Knowledge-Creating Company: How Japanese Companies Create the Dynamics of Innovation. Oxford University Press, Oxford (1995)

# Segmentation of Heart Image Sequences Based on Human Way of Recognition

Arkadiusz Tomczyk and Piotr S. Szczepaniak

Institute of Computer Science, Technical University of Lodz,
Wolczanska 215, 90-924 Lodz, Poland
tomczyk@ics.p.lodz.pl

**Abstract.** The paper presents a method of heart ventricle segmentation. The proposed approach tries to imitate a human process of top-down localisation of proper contour by subsequent detection of a whole heart, interventricular septum and, finally, left and right ventricle. As a tool of cotour detection adaptive potential active contours (APAC) are used. They allow both to describe smooth, medical shapes and, like the other active contour techniques, make it possible to use any expert knowledge in energy function. Finally, the paper discusses the obtained results.

## 1 Introduction

Image segmentation is a crucial element of every system where imitation of human image recognition process is necessary. There exist numerous methods of low-level segmentation such as thresholding, edge-based methods, region-based methods, etc ([1,2,3]). Most of them, however, possess limitations when more complicated images (e.g. medical images) are considered. Those limitations are a result of fact that proper segmentation in such cases is possible only if additional, high-level expert knowledge is considered. In this paper this knowledge is used in three-phase top-down process of left and righ heart ventricle segmentation. As a tool, which does not possess limitations mentioned above, adaptive potential active contours ([4,5]) are used. They, simirarly to other active contour techniques ([6,7,8]), allow to incoporate any expert knowledge in the energy function.

The paper is organized as follows: in section 2 the considered problem is presented, section 3 describes proposed three-phase approach to heart ventricle segmentation, section 4 is devoted to discussion and presentation of obtained results, whereas the last section focuses on the summary of the proposed approach.

## 2 Problem

Obstruction of pulmonary arteries by the emboli is one of the most frequent cause of death among the population in the most developed societies. In recent

N. Zhong et al. (Eds.): BI 2009, LNAI 5819, pp. 225–235, 2009.

**Fig. 1.** Sample heart image: (a) - sample slice, (b) - masks of right ventricle obtained from contours drawn by an expert ($M^{ri,in}$ - white, $M^{ri,ou}$ - black), (c) - masks of left ventricle obtained from contours drawn by an expert ($M^{le,in}$ - white, $M^{le,ou}$ - black)

decades the development of computed tomography (CT) allowed to introduce new diagnostic methods of pulmonary embolism. Some of them were mentioned in [4]. This paper focuses on a method for the right ventricle systolic pressure assessment linked with shifts in interventricular septum curvature that requires drawing of endocardial and epicardial contours. Currently they are drawn manually, which is time-consuming as well as prone to errors due to technical mistakes and tiredness (Fig. 1). The aim of this work is to automate this process.

The analysed image data are obtained using ECG-gated computed tomography scanner in a standard chest protocol after intravenous injection of contrast media. Next, heart cycle is reconstructed in 10 phases and two chamber short axis view is generated leading to 4D image sequence, i.e. the set of 2D images for each reconstructed slice and for each phase of heart cycle.

## 3   Proposed Approach

The proposed method tries automatically to imitate the top-down process of right and left heart ventricle contour localization. This process is divided into three phases, i.e. preprocessing, segmentation and improvement phase that are described below. It uses adaptive potential active contours (APAC) presented in [4,5] with different formulation of energy function in each phase. The choice of potential active contour approach resulted from specific properties of potential contour model that, in a natural way, allows to describe smooth, medical shapes. Other traditional active contour techniques such as snakes or geometric active contours require special steps, e.g. definition of additional smoothing internal energy components, in order to achieve similiar results. It should be emphasized that the presented approach is currently not fully automatic because the expert must choose, for each 4D image sequence, a proper threshold that allows to distinguish in the images those areas that represent blood with injected contrast from the rest of the image (Fig. 2a). This requirement, however, in comparison with manual contour drawing is not a difficult task. Nevertheless, there are also conducted works that aim at solving that problem. Below the following denotations are used:

– $I$ - set of all pixels in the image;

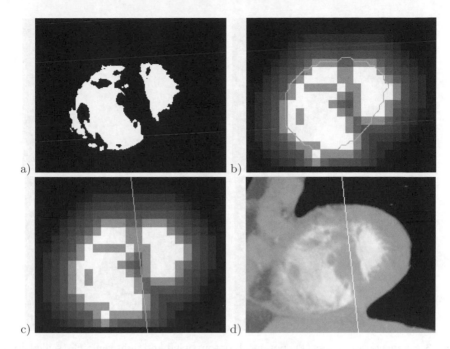

**Fig. 2.** Preprocessing phase: (a) - image after thresholding, (b) - potential contour of both ventricles (scaled down image after thresholding with weak distance potential filter), (c) - interventricular septum line (scaled down image after thresholding with weak distance potential filter) (d) - interventricular septum (original image)

- $C^{in}$, $C^{ou}$ - set of pixels in the image that lay inside and outside the current contour, respectively;
- $L^{ri}$, $L^{le}$ - set of pixels in the image that lay on the right and left side of the current line separating ventricels, respectively;
- $M^{ri,in}$, $M^{ri,ou}$ - set of pixels in the image that lay inside and outside the right ventricle mask drawn by an expert, respectively;
- $M^{le,in}$, $M^{le,ou}$ - set of pixels in the image that lay inside and outside the left ventricle mask drawn by an expert, respectively;
- $T_k^{in}$, $T_k^{ou}$ - set of pixels in the image that lay inside the rectangle drawn around the $k$-th contour point and lay inside and outside the current contour, respectively.
- $n$ - function that value represents the number of pixels in a given set;
- $D$ - function that value represents the intensity at the given point of a processed image (the values are in range $[0, 1]$) and which definition differs between phases;

### 3.1   Preprocessing Phase

The first element of the proposed top-down detection process is localisation of both heart ventricles. In this phase potential active contours are used. At this

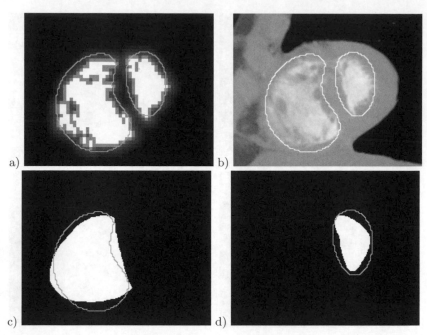

**Fig. 3.** Segmentation phase: (a) - potential contour of left and right ventricle (scaled down image after thresholding with weak distance potential filter), potential contour of left and right ventricle (original image), (c) - potential contour of right ventricle and expert mask ($M^{ri,in}$ - white), (d) - potential contour of left ventricle and expert mask ($M^{le,in}$ - white)

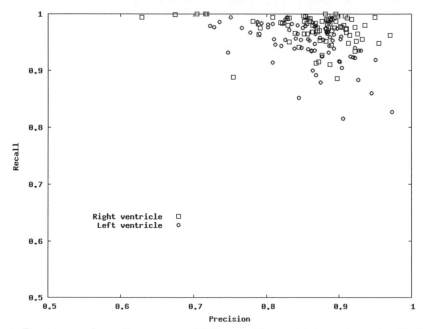

**Fig. 4.** Precision and recall measures of left and right ventricle contours for 80 slices of the same heart

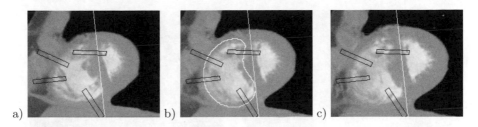

**Fig. 5.** Improvement energy component, rectangles represent 4 sample areas where local avarage pixel intensity inside and outside the contour was considered: (a) - slice below current segmented image, (b) - current segmented image, (c) - slice above current segmented image

stage energy function focuses on two aspects of the contour that is sought: it should be relatively small and should contain inside all the pixels representing blood with injected contrast (pixels with intensity above the given threshold) (Fig. 2a). Consequently, energy has the bigger value the larger contour is and the more pixels above threshold stay outside the contour:

$$E^{pre} = \frac{n(C^{in})}{n(I)} + 4\frac{\sum_{\mathbf{x} \in C^{ou}} D(\mathbf{x})}{n(C^{ou})} \tag{1}$$

Here $D$ is a strongly scaled down image after thresholding with weak distance potential filter applied (Fig. 2b, Fig. 2c). The weights of the energy components are chosen experimentally. Scaling down of the image reduces time of computations and distance potential filter causes better contour behaviour during evolution (there is no problem with locally constant character of objective function).

Detection of both ventricles allows to localize approximatelly interventricular septum because it should lay inside that contour and should have significantly lower intensity than pixels representing blood with contrast. In the presented approach septum is represented by a line separating left and right ventricle. This line is found using brute force search algorithm that tests each line starting at the top edge of the image and ends at its bottom edge. This line is accepted that has the longest part inside the prevoisuly found contour and has, at that part, the maximum amount of points below the given threshold (Fig. 2d). In fact, the search of this line can also be considered as an another active contour algorithm where evolution method evaluates every possible contour to find the optimal one.

### 3.2    Segmentation Phase

Having found the line representing interventricular septum, it is possible to find a left and right ventricle using the similar approach described in preprocessing phase. Here also the potential active contour can be of use. In order to find the ventricle energy function one has to take into consideration three aspectes of the sought contour: it should be relatively small, it shold lay on the proper side of the septum line and it should contain inside all the pixels representing blood

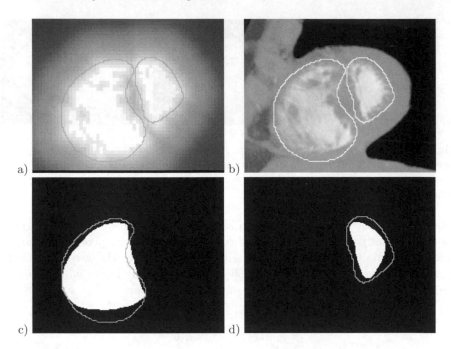

**Fig. 6.** Improvement phase: (a) - potential contour of left and right ventricle (scaled down image after thresholding with weak distance potential filter), potential contour of left and right ventricle (original image), (c) - potential contour of right ventricle and expert mask ($M^{ri,in}$ - white), (d) - potential contour of left ventricle and expert mask ($M^{le,in}$ - white)

with injected contrast that lay on the proper side of the septum line. To put it differently, for the right ventricle energy has the bigger value the larger contour is, the more pixels stays at the right side of the septum line and the more pixels above threshold stay outside the contour at the right side of that line:

$$E^{seg,ri} = 2\frac{n(C^{in})}{n(I)} + \frac{n(L^{ri})}{n(I)} + 4\frac{\sum_{\mathbf{x}\in C^{ou}\cap L^{le}} D(\mathbf{x})}{n(C^{ou}\cap L^{le})} \tag{2}$$

Similarily, for the left ventricle energy has the bigger value the larger contour is, the more pixels stays at the left side of the septum line and the more pixels above threshold stay outside the contour at the left side of that line:

$$E^{seg,le} = 2\frac{n(C^{in})}{n(I)} + \frac{n(L^{le})}{n(I)} + 4\frac{\sum_{\mathbf{x}\in C^{ou}\cap L^{ri}} D(\mathbf{x})}{n(C^{ou}\cap L^{ri})} \tag{3}$$

Here $D$ is a slightly scaled down image after thresholding with weak distance potential filter applied (Fig. 3a). The weights of the energy components are chosen experimentally.

## 3.3   Improvement Phase

The segmentation phase gives quite well results but is designed only to find just approximate localization of both ventricles. It does not take into consideration any additional expert knowledge that should be considered (more detailed description of that need will be discussed further). In this work a sample improvement is used which shoud imitate the behaviour of expert that manually draws the contours. As it was mentioned earlier, 2D images are the result of reconstruction, which aims at creating two chamber short axis view. Thus, the pixel intensities of the obtained slices are averaged. That is why during image examination, in order to find a proper contour, an expert must look not only at the currently segmented slice bu also at the slices above and below. This is imitated using the following energy forumulation for the right ventricle:

$$E^{imp,ri} = \frac{\sum_{\mathbf{x} \in C^{ou} \cap L^{le}} D(\mathbf{x})}{n(C^{ou} \cap L^{le})} + O^{imp,ri} \tag{4}$$

and for left ventricle:

$$E^{imp,le} = \frac{\sum_{\mathbf{x} \in C^{ou} \cap L^{ri}} D(\mathbf{x})}{n(C^{ou} \cap L^{ri})} + O^{imp,le} \tag{5}$$

Here $D$ is a slightly scaled down image after thresholding with a strong distance potential filter applied (Fig. 6a). The weights of the energy components (equal to 1) are chosen experimentally.

The first component of the above equations ensures that all the pixels representing blood with injected contrast that lay on the proper side of the septum line are inside the contour. The second component (identical for both ventricles which need not to be a rule) is used to make the contour, in the given $K$ points equally distributed along the contour, to lay there where local differences between avarage pixel intensities (from the current slice and the slices above and below) outside and inside the contour are the biggest:

$$O^{imp,ri} = O^{imp,le} - \frac{1}{K} \sum_{k=1}^{K} (1 - u_k) \tag{6}$$

where

$$u_k = \begin{cases} t_k & \text{if } t_k \geq 0 \\ 0 & \text{if } t_k < 0 \end{cases} \tag{7}$$

and

$$t_k = \frac{\sum_{\mathbf{x} \in T_k^{in}} D(\mathbf{x})}{n(T_k^{in})} - \frac{\sum_{\mathbf{x} \in T_k^{ou}} D(\mathbf{x}))}{n(T_k^{ou})} \tag{8}$$

Here $D$ is a an image obtained by avaraging (after thresholding) the pixels of the current slice and of the slices that lay directly below and above the current slice (Fig. 5).

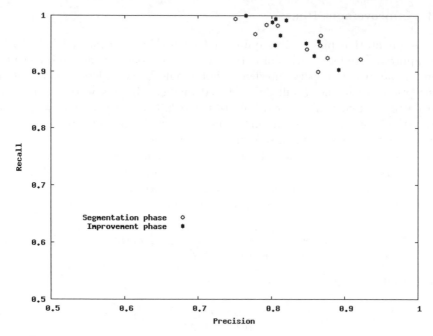

**Fig. 7.** Comparison of precision and recall measures of left ventricle contours in segmentation and improvement phases for 10 slices of the same heart during 1 heart cycle

## 4    Discussion of the Results

The conducted exepriments revealed that a preprocessing phase is able to find a proper septum line if only a proper theshold is chosen by an expert (Fig. 2c, Fig. 2d). Also, the segmentation phase gives satisfactory results (Fig. 3) though in this case an additional objective analysis should be performed to evaluate the obtained contours (approximate septum line can be evaluated visually).

As shown in [9] contours can be treated as binary classifiers of pixels. As a result, so as to evaluate them methods of classifier evaluation can be used. In this paper precision $P$ and recall $R$ are used for this purpose. For the right ventricle they can be expressed as:

$$P^{ri} = \frac{n(M^{ri,in} \cap C^{in})}{n(M^{ri,in} \cap C^{in}) + n(M^{ri,ou} \cap C^{in})} \tag{9}$$

$$R^{ri} = \frac{n(M^{ri,in} \cap C^{in})}{n(M^{ri,in} \cap C^{in}) + n(M^{ri,in} \cap C^{ou})} \tag{10}$$

and for left ventricle as:

$$P^{le} = \frac{n(M^{le,in} \cap C^{in})}{n(M^{le,in} \cap C^{in}) + n(M^{le,ou} \cap C^{in})} \tag{11}$$

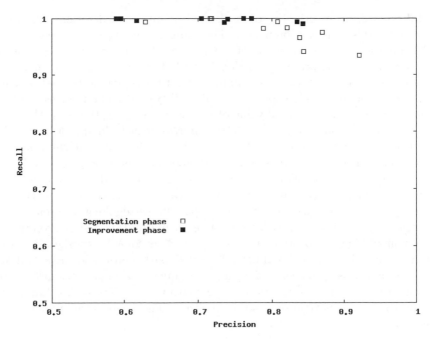

**Fig. 8.** Comparison of precision and recall measures of right ventricle contours in segmentation and improvement phases for 10 slices of the same heart during 1 heart cycle

$$R^{le} = \frac{n(M^{le,in} \cap C^{in})}{n(M^{le,in} \cap C^{in}) + n(M^{le,in} \cap C^{ou})} \tag{12}$$

Precision indicates how many pixels inside the current contour are in fact correctly classified and recall how many pixels that are expected to be classified are in fact inside the contour. The more both of those measures at the same time are closer to 1, the more perfect segmentation results are. Smaller values of precision mean that contour sticks out of the mask drawn by an expert and smaller values of recall mean that mask sticks out of the contour. Those measures are in general not well suited for segmentation problem because they consider only a number of pixel (contour area) and not its structure (contour shape) but for the purposes of this article they are sufficient.

The above measures were used to evaluate the segmentation phase results. The results for 80 (8 slices and 10 phases of heart cycle) 2D images and both left and right ventricle are presented in Fig. 4. For most of them the precision and recall have simultaneously values above 0.8 which is a quite satisfying result.

The visual effect of segmentation phase for a sample slice is presented in Fig. 3. It seems to be quite well if the orginal image is considered (Fig. 3b). It is quite well fitted to the image data but in comparison to the masks drawn by an expert (Fig. 3c, Fig. 3d) it differs significantly. The reason of that is a fact that an expert uses additional knowledge during segmentation. The improvement phase described above focuses on one technical aspect of manual segmentation where the information from other slices is also taken into consideration. The initial results after this phase are promising. The comparision of the contours obtained for 10 (1 chosen slice and 10 phases of heart cycle) 2D images in segmentation and improvement phase is presented in Fig. 7 and Fig. 8 (the reslut of segmentation phase was the initial contour in the improvement phase and additional potential sources were added to increase contour flexibility). A slight improvement can be observed especially in recall value of the right ventricle. However, the visual effect (Fig. 6c, Fig 6d) reveals that better improvement technique still should be searched for. The reason behind it is a fact that only visual information (even if additional slices are used) is considered and an expert possesses additional medical knowledge that allows to draw a proper contour. It can be used in the presented framework because any information can be used in energy function. For example, one can consider the conscious expert knowledge (e.g. the relationship between shapes of both ventricles such as information that the interventricular septum should have a constant thickness) as well as some knowledge that is not conscious or is hard to express directly using mathematical formula and that can be obtained using machine learning techniques (e.g. neural network trained to estimate the proper localization of the contour in the given slice taking into account the image information from other slices). These aspects are under further investigation.

## 5   Summary

In this paper three-phase approach for segmentation of left and right heart ventricle was presented. The approach in question imitates the conscious top-down human process of segmentation using adaptive potential active contour (APAC) technique. The obtained results reveal that proposed methodology is well suited to the described problem. The preprocessing and segmentation phases allow to achive very good approximate localization of heart ventricles. However, a precise review of expert results requires additional semantical knowledge. The proposed improvement phase uses only visual information and, thus, the the improvement is only slight. It shows that in that phase other formulations of energy function, especially those that are able to use unconscious or hard to express medical expert knowledge, should be used, which will be further investigated by the authors of this work.

## Acknowledgement

This work has been supported by the Ministry of Science and Higher Education, Republic of Poland, under project number N 519007 32/0978; decision no. 0978/T02/2007/32.

Authors would like to express their gratitude to Mr Cyprian Wolski, MD, from the Department of Radiology and Diagnostic Imaging of Barlicki University Hospital in Lodz for making heart images available and sharing his medical knowledge.

# References

1. Tadeusiewicz, R., Flasinski, M.: Rozpoznawanie obrazow. Wydawnictwo Naukowe PWN, Warszawa (1991) (in Polish)
2. Sonka, M., Hlavec, V., Boyle, R.: Image Processing, Analysis and Machine Vision. Chapman and Hall, Cambridge (1994)
3. Gonzalez, R., Woods, R.R.: Digital Image Processing. Prentice-Hall Inc., New Jersey (2002)
4. Tomczyk, A., Wolski, C., Szczepaniak, P.S., Rotkiewicz, A.: Analysis of Changes in Heart Ventricle Shape Using Contextual Potential Active Contour. In: Proceedings of 6th International Conference on Computer Recognition Systems CORES 2009 (accepted for publication)
5. Tomczyk, A.: Image segmentation using adaptive potential active contours. In: Kurzynski, M., Puchala, E., Wozniak, M., Zolnierek, A. (eds.) Proceedings of 5th International Conference on Computer Recognition Systems CORES 2007, Adavnces in Soft Computing, pp. 148–155. Springer, Heidelberg (2007)
6. Kass, M., Witkin, W., Terzopoulos, S.: Snakes: Active contour models. International Journal of Computer Vision 1(4), 321–333 (1988)
7. Caselles, V., Kimmel, R., Sapiro, G.: Geodesic active contours. International Journal of Computer Vision 22(1), 61–79 (2000)
8. Grzeszczuk, R., Levin, D.: Brownian strings: Segmenting images with stochastically deformable models. IEEE Transactions on Pattern Analysis and Machine Intelligence 19(10), 1100–1113 (1997)
9. Tomczyk, A., Szczepaniak, P.S.: On the relationship between active contours and contextual classification. In: Kurzynski, M., et al. (eds.) Proceedings of the 4th Int. Conference on Computer Recognition Systems CORES 2005, pp. 303–310. Springer, Heidelberg (2005)

# Author Index

Printing: Mercedes-Druck, Berlin
Binding: Stein+Lehmann, Berlin